SMART MOUTH

SMART MOUTH

WINE ESSENTIALS FOR YOU, ME, AND EVERYONE WE KNOW

JORDAN SALCITO

ILLUSTRATIONS BY JENNY BOWERS

TEN SPEED PRESS
California | New York

FOR MY PARENTS,
DON AND CATHERINE,
AND FOR ROBERT

CONTENTS

INTRODUCTION

HOW I FELL IN LOVE WITH WINE

Three months after college, my parents helped me move into a small window-less room above the headquarters of the Hell's Angels. It was August in the East Village, and the block smelled like garbage and stale beer—summer in New York. A cockroach scuttled along the wall of the elevator shaft. My parents stifled tears while my roommate, an acquaintance from college, said I should be glad we had an elevator, as most first New York apartments were walk-ups.

When I moved into that room I had a few hundred dollars, a degree in English literature, and a desire to write. My parents had hoped I might follow my father's footsteps to law school but had encouraged me to pursue something I loved as a career. I was still figuring out what that meant.

I spent the first week in my new city delivering resumes to restaurants in the neighborhood. Until landing the writing job of my dreams, I needed to figure out how to pay rent. Of the twenty places I visited, nineteen did not call me back. The twentieth, on the corner of First Avenue and Twelfth Street, was a restaurant named Tappo (the Italian word for "cork"). Tappo had a large wooden bar and exposed brick walls lined with cabinets cross-checked in diamond shapes, filled with wine bottles lying horizontally, collecting dust. The kitchen was run by a talented man named Saul who had trained there with the famous chef Jodi Williams before she left. The general manager, Ron, hired me. He was my height, five foot seven, with brown hair, an ill-fitting gray suit, and a mustache like Tom Selleck's in *Magnum, P.I.* He seemed to belong in Midtown, or another town entirely. Ron

was my ticket to the "New York restaurant experience," an important line item for anyone seeking restaurant employment in Manhattan. He offered me a hostess position, and I started the next day.

Soon that hostess job turned into a management role and, in an attempt to increase sales and keep their sinking restaurant afloat, the owners inaugurated internal wine tastings on Thursday afternoons. I don't recall the particular bottles we tried, only that most were European and more interesting than anything I'd tasted up to that point. My wine education until then had involved sipping Chianti and Châteauneuf-du-Pape from my parents' glasses at dinner tables and drinking cheap Chianti and Prosecco from a straw during my college semester abroad. A wine connoisseur I was not. There were many elements of that restaurant that did not feel like a good fit, but I was drawn to it anyway because of those wines.

The underlying purpose of these Thursday tastings seemed to be an opportunity for the owners—an unmarried couple in a tumultuous relationship—to communicate their feelings for each other. If they were compatible on tasting day, they found the wines "generous" and "sensual." If they were arguing, the wines were "angry" and "closed." In retrospect, this unconventional introduction to wine descriptions served as an interesting and ultimately beneficial lesson in the way we, as individuals, experience what we consume. It was at Tappo that I began to develop a language for discussing wines with restaurant guests; inexplicably, it was a way that resonated with them. A gorgeous Spanish Rioja reminded me of Penélope Cruz: voluptuous, striking (to my palate at the time), and full of beauty. A Sauvignon Blanc from Sancerre reminded me of Gwyneth Paltrow: lithe and steely, full of resolve.

My language for wine has evolved significantly since then. But as a young postgrad, these personas made wine easy to connect with, and they helped me make sense of the different expressions of regions and grapes. That eccentric wine education helped clarify that what speaks to us in a wine is intuitive and personal. Even after years of exams and analytical training, it's more clear to me than ever that the wine descriptions that matter are ones that resonate with *you*.

I left Tappo to work at a restaurant called wd~50, an iconic molecular gastronomy restaurant on Clinton Street on the Lower East Side, run by the chef Wylie Dufresne. Wylie valued curiosity and research and included anyone who showed signs of interest in menu tastings and conversations about meat glue, methylcellulose, and "soil" made of peas. It was at wd~50 that I began to consider a life in restaurants and redirected my dreams.

Eighteen months later, with elation, impostor syndrome, and a culinary degree from Johnson & Wales University, I moved to the Upper East Side to work in the kitchen at Daniel Boulud's flagship restaurant, Daniel. In culinary school I'd read his book *Letters to a Young Chef* and had dreamed of someday working there. I began my job as a kitchen stagiaire (intern), and it would only be a few more months before I fell fully in love with wine.

Long before my time at Daniel, wine had already assumed an invisible pull, a connective and exhilarating quality I was not yet able to articulate. My father's father, an Italian immigrant who died when my dad was thirteen, made wine in his basement in Waterbury, Connecticut. Making wine together is the only memory my father ever shares of his dad. When I was young, I tried sips of my parents' wines at dinner, but they did not strike me as particularly special—just fermented grape juice, offering a promise to a world of mystery and romance—without that mystery or romance realized.

By the end of my six-month kitchen internship, my North Star had shifted. The change began gradually, with bottles I tasted during nights off. Then it was spurred by an impromptu conversation in the kitchen with Daniel Boulud himself over a glass of 1990 Domaine Jean-Louis Chave Hermitage Rouge. I had worked the garde-manger station and spent much of that evening wrapping fish fillets in potato strips for the restaurant's signature dish at the time, the Black Bass en Paupiette. Before heading home, I'd swung by the kitchen to snack on abandoned pastries and happened to be carrying an out-of-print cookbook called *Great French Chefs*. Moments later, with seemingly divine timing, Daniel walked into the kitchen, holding a decanter.

"Hallo," he said, eyeing me, then the book. "What do you do in my kitchen?" Followed by, "And, do you like wine?" He placed the decanter onto the pass and poured me a small glass while I handed him the book. The wine, made of Syrah and grown on the famed hill of Hermitage, in the Northern Rhône, tasted of bacon and leather, tobacco and licorice, black peppercorns, blackberries, and smoke. It was the first time I tasted a Syrah that made my heart stop. When Daniel landed on the page highlighting legendary chef Paul Bocuse's recipe for Red Mullet with Potato Scales, his eyes lit like sparklers. Daniel grew reflective and explained he had flown to Lyon to gain inspiration at Restaurant Paul Bocuse between leaving his role helming the kitchen at Le Cirque and opening his own flagship restaurant, Daniel, in New York. That red mullet dish, shimmering up at us from the page on the pass, had inspired the very Paupiette I helped prepare every night. To further the synchronicity, the garnish for the dish in the book was a simple but striking sauce, velvety and rich, made of butter, shallots, fresh thyme, and Northern Rhône Syrah.

During my first summer of employment at Daniel, I began dating Robert, a man with whom I had been gradually falling in love with for a year and whom I would eventually marry. He was tall and handsome, with a sense of humor sharpened in Parsippany, New Jersey, an encyclopedic understanding of Burgundy, and a closet full of Kiton ties. Robert, the first in his family to attend college, had turned down admission to Harvard Law School to continue working in restaurants, something his mentor, John Sexton, the dean of NYU Law School at the time, told him he should do until he no longer loved doing it. When we began dating, Robert was the sommelier and partner at a restaurant called Cru on lower Fifth Avenue, near Washington Square Park. Cru was fancy but warm. It had large windows that opened onto Fifth Avenue and a cocobolo wood bar, which looked like milk and dark chocolate swirled together. The wine list was so large it required two leather-bound volumes—separate books for red and white wines, with hand-drawn maps printed on pages woven with linen. Most impressive was the wine collection Robert had assembled.

By autumn, we saw each other regularly. While that sometimes involved late-night dinners at places like Blue Ribbon (where we ate roasted bone marrow with brioche and drank underpriced bottles of Domaine Roulot Meursault), evenings often began at Cru. I'd sit at the bar and Robert would open bottles from his favorite producers, which he'd pour for me "blind," without showing me the label. On one occasion, he opened a bottle of 1980 Henri Jayer Vosne-Romanée Premier Cru Les Brûlées "for research," he said, before pouring a ribbon of wine slowly into an enormous glass that looked like a fishbowl. It captured smells of a faraway time and place, which I could access over and over again upon sticking my nose into the glass. I still have the label, pressed between the pages of a Michel Bras cookbook.

Robert introduced me to wines that brought me back in time, that tasted of history and reverence and tart cherries with hints of green tobacco and limestone. Our timing was good, as far as wine education goes. We met in 2003, began dating in 2005, and married in 2008 two weeks before Lehman Brothers collapsed. That was the same year Hong Kong removed its 80 percent wine importation tax, taking it first to 40 percent and then to zero. Hong Kong's market demand for wine skyrocketed, and China's mainland soon followed. Today, a bottle of 1980 Jayer Les Brûlées trades for around eleven thousand dollars—if you can find one. It was still a splurge prior to 2008, but at a few hundred dollars, it cost significantly less at that time.

On nights we didn't go out, I'd get home from work and stay up until well past midnight reading Remington Norman's *The Great Domaines of Burgundy* and Clive Coates's *Côte d'Or.* I wanted to understand what and who made these wines that stirred my soul. (Jasper Morris's seminal *Inside Burgundy* had not been written yet, but if you love Burgundy, that's a must for your library.)

My life changed in 2006 at The Little Nell ski resort in Aspen, Colorado, during a Burgundian harvest celebration called La Paulée. This was the first year the event was held outside of California or New York. It was also smaller and more intimate than the usual programming. Knowing with some sort of higher intelligence that I needed to attend, I asked to work at the event for free and was granted admission. It was at La Paulée I met several people who would become instrumental mentors in my journey and career, including winemakers, sommeliers, collectors, and chefs. That was my aha moment, realizing wine is, at its highest purpose, an extraordinary connective tissue and a portal to a life of adventure and love.

At the final meal in Aspen, I sautéed chicken oysters and chanterelles alongside Daniel Boulud, applying more focus and intention to this task than I had ever applied to anything in my life. When the last dish left the pass for the dining room, Daniel took me aside. "It is clear you love wine, Giordanne," he said, elongating the last syllable of my name in a way that made it sound French, and more beautiful. "You can always have a job in my kitchen, but that is not where your heart is." My next step, he suggested, was to work the harvest in France.

Nine months later, in September, I was wearing rubber overalls and borrowed Wellingtons, picking grapes in a muddy Clos des Fôrets vineyard in Burgundy's Côte de Nuits. There I began a lifelong relationship with the region that opened my eyes, and the door, to a career in wine. Eighteen months after that, I was dancing to Journey's "Don't Stop Believin' " on the dining room credenza at Eleven Madison Park, drinking Roger Pouillon Champagne from the bottle with my colleagues, celebrating our James Beard Award for Outstanding Wine Service under the direction of my boss, John Ragan.

Eight years after that, while taking a break from restaurants, I worked a wine event in Healdsburg, California, and reconnected with David Chang, who offered me an opportunity that felt like a dream: overseeing the beverage programs for his restaurants throughout the United States. Dave's mandate was exhilarating. He said, "Put Momofuku on the map as a great place for wine," followed by, "You know the rules. Now break them." Nothing unleashes creative potential quite like that sort of directive. I spent the first few months eliminating 750-ml bottles from Noodle Bar's wine program, replacing them with half bottles and magnums instead, then building out the wine cellar at Ma Pêche, celebrating great wines of the world, along with a section we called 56 Selections: off-the-beaten-path bottles sold for fifty-six dollars each. Soon after, we turned the pairing menu at Ko into an all-sparkling affair. This included a never-fail pairing born out of near despair—a wineglass rinsed with vin jaune, topped with Crémant de Jura. Oxidative wines like vin jaune and sherry are very complex on a molecular level, and this rinse element was a trick we kept in our back pockets for pairings spanning anything from shishito pepper consommé to uni with Momofuku's signature umami chickpea purée. The pairing grew out of both a rigid commitment to staying within the confines of an all-sparkling pairing and the creative tug-of-war that ensues when working with chefs who are keen to test your abilities and have not yet discovered they love wine.

We built a list I was proud of, focusing on visionaries who broke rules in the spirit of excellence. Our list celebrated the wines of established rule-breakers like Anselme Selosse, who changed the perception and direction of excellence in Champagne and influenced the following generation of winemakers. To help forge a connection with the bottles on hand, I added photographs taken in wineries over the years. The James Beard Foundation included Ko as a semifinalist for Outstanding Wine Service, and we took home the award for Most Creative Wine List in the World from *World of Fine Wine* magazine.

Around this time, I stood on the precipice of passing the notoriously difficult three-part Master Sommelier exam. I had poured myself into study so completely—from memorizing minutiae and drawing maps to daily mock exams and blind tastings—that the exam felt like a breeze. I nailed Tasting and Theory and sailed through Service, answering questions correctly and having fun all the while. After the exam, I learned I'd failed the Service Practical and reflected on the feedback I'd received. A table of judges who neither knew me nor had seen me work in a restaurant—nor worked in restaurants themselves—failed me for "not seeming like myself" on a day when I had never felt more like myself nor more confident in my ability to provide exceptional service. This all occurred in a moment of my life when I had never been more fluent in wine service, when I had never been prouder of or had received more national acclaim for the wine program I built and oversaw. The test outcome was confounding. I felt gratitude for the process and my own journey. I also felt disappointed and powerless. Then I began to question everything I had been working toward.

A week later, at a wedding in Italy, I learned I was pregnant. That was the change agent I had not realized I'd needed that ultimately led me to dust off the idea for RAMONA, my organic Italian spritz company, from deep within my subconscious and create what would become my life's work (along with, of course, being a mother). As I've begun to realize, when you give something your all and it doesn't work out, the universe has something better in mind.

This book channels that spirit. Here I provide a framework for learning about wine, influenced by the structure and rigor of exams, information learned through experiences, and a healthy disregard for authority—or entrenched institutions—when I believe the motives are questionable. (Soon after my exam, the Court of Master Sommeliers became embroiled in several national scandals ranging from alleged systemic sexism and sexual harassment to a widespread cheating debacle.) My hope is that this book arms you with information or inspires you to ask questions that will allow you to discover what you like, find out what moves you, and learn for yourself what resonates. In wine, as with anything, knowledge is power.

MY BURGUNDIAN BEGINNINGS

Burgundy is as close to religion as anything I've encountered in life. (If you are thinking to yourself, *I love Burgundies but cannot afford to drink them,* then fear not! There are still spectacular producers under the radar—a few of my favorites are in the "Producers" section in the appendix on page 231.) I fell hard for these ethereal expressions of Pinot Noir and Chardonnay, each a liquid time capsule representing a family or vineyard history often dating back hundreds of years. I fell for the region and its vignerons (aka winemakers, although the direct translation here is "wine grower," as there is no French word for "winemaker"), for their humility, vision, commitment to excellence, and respect for their history and their land.

Burgundy served as a glorious gateway to a life in wine. Working a harvest that first vintage lit a fire in me to return as often as possible. Robert and I began a decade-long tradition, flying to France roughly one hundred days after budbreak to assist harvest at a different domaine each year, building upon the knowledge from previous years with each new visit. After we finished our work in the winery I'd run through the vineyards (I was still an avid runner in those days), collecting pieces of limestone from vineyards with poetic names—Musigny, Reignots, and La Tâche, some of the wine world's most storied plots of land. At dinner we would blind taste and discuss wines from the cellar, my first—and very informal—exposure to the virtues of blind tasting wine alongside people who actually had informed opinions.

What became evident during those harvests is that excellence in wine—and in anything—is a result of someone who cares a great deal and consistently chooses quality at every turn. Great winemakers are also artists and visionaries; their medium is fermented grapes. While some of these wines can get quite expensive, many are not, once you know what to look for—and many of the pricier ones are arguably worth the splurge.

Wine is forever a journey. And no, you do not need to work harvest in France to appreciate it. Everyone's palate differs, as does everyone's starting place. The average human has between two thousand and ten thousand taste buds, all configured differently. No two people have the same number or orientation: your tongue is as unique as your fingerprint, and it shape-shifts all the time. Add to that your own unique set of life experiences that will invariably inform your interests and, therefore, your path.

HOW TO USE THIS BOOK

When this book was just an idea, I asked a table of friends what they wanted in a wine book and one of them said, "Please write a book that takes me from a D student to an A." Picking up a book is an investment of time, your most valuable resource. I'm going to assume you are interested in being as smart as you can be about wine, as efficiently as possible. This book is for drinkers who are thoughtful and curious, who want to understand why they love the wines they do, and want to learn how to replicate that experience and share it with friends. It is meant to equip you with the tools you need to feel confident in any wine-related conversation and to provide questions for consideration along the way. It is meant to offer you flexibility. Read it chronologically if you'd like, or dip in and out of whatever section is most interesting or relevant to you at a given time. There is no "right" or "wrong" way to learn about wine. Start where you are, and start when you're ready. You're in the driver's seat.

In that spirit, I've decided to dive right into the "Tasting" section at the beginning, because I believe that is the most natural starting point. You are very likely reading this book because you tasted a bottle that captured your attention, and you want to know more. There are no shortcuts to learning about wine, but one tried-and-true route is through focused tasting—with people who are practiced at it, if you can find them. Tasting is a terrific way to understand the mechanics of what's in a glass: the quality of the grapes, the climate and soil in which they grew, how they were vinified (made into wine), the winemaker's philosophy—and of course, to decide if you like it. The tasting section will likely spark a lot of questions. Don't fear! My goal is to peel back the curtain through this first section, then answer the questions that might arise more in depth through the rest of the book. Throughout, you'll also find sidebars (which I think of as "Extra Pours")—these provide history or color on topics that have captured my curiosity and that I believe may also spark yours. At the very least, they will prove useful nuggets to sprinkle into conversation at dinner parties.

In chapter 1 we begin with "Tasting," thanks to a heavy dose of inspiration from Jay Fletcher, a Master Sommelier based in Aspen, Colorado, who is a terrific taster, teacher, and human. We'll focus on what you can learn from the appearance, smell, taste, and structure of a wine, and how that multisensory experience can make you a more knowledgeable, smarter consumer. While you don't need to learn to blind taste to appreciate wine, it's a great way to objectively understand quality. There's a lot of logic in every bottle, and developing a knack for analytical tasting will help crystallize that.

Chapters 2, 3, and 4 explore an overview of wine styles, the basics of winemaking, and the environment, respectively. We'll unpack how to craft wines that are white, red, orange, or pink; sparkling or still; and dry or sweet. We'll also cover what *natural* means and doesn't, philosophically as well as in actual concrete, regulated terms. We'll discuss land and environment on both a macro and a micro level, from a vineyard's location to the health of microbiomes (tiny fungi and healthy bacteria) in the soil. We'll also dive into topics like climates, rocks and soils, farming practices and terroir, and how each of these factors influences what ends up in your glass. You'll find an ode to maps here, too.

Chapter 5 focuses on grapes you're most likely to encounter in restaurants and wine lists. This chapter is meant as an overview and a starting point; there are more than ten thousand known wine grapes in the world! Additionally, I've added a cheat sheet to help make it easier to understand which grapes are used in which regions, so it's less of a mystery.

Chapter 6 is an overview of "Old World" countries and regions, defined as nations who were cultivating *Vitis vinifera* grapes prior to the fourteenth century. Throughout the vast majority of wine's six-thousand-year history, grape varieties and regions (on account of climate and soil) are two parts of a whole. This chapter examines wine through the lens of the geography, culture, and national identity they're part of and influenced by today. When you purchase a bottle of white wine labeled "Sancerre," for example, you are buying a bottle made from Sauvignon Blanc grapes, grown in the Loire Valley in France. (While we're here, Sancerre is also a region known for terrific red and rosé wines, which are always made from 100 percent Pinot Noir.) That bottle is an expression of a winemaker's point of view, grapes grown in a specific place, and a labeling system determined by

France, within the larger construct of the European Union. As you can imagine, every word on a wine bottle is regulated.

Chapter 7 explores "New World" countries and regions, referring geographically to wine regions developed during and after the fifteenth century as a result of European exploration and colonization. Since the mid-nineteenth century, "New World" countries have also tended to embrace technology over tradition, even though these lines have begun to blur. A bottle of Pinot Noir from the Willamette Valley in Oregon or Central Otago in New Zealand may both involve the same grape, but soil types, climate, and the country's own labeling system influence the taste of that bottle, as does the winemaker's intention and point of view.

Chapter 8 examines a wine's journey to you. Here, we'll cover the world of distribution—the trip a bottle takes before it arrives at restaurants, wine shops, and hotels. If you've ever wondered why you see a lot of the same mediocre wines everywhere, even when you feel you deserve better options (you do!), we'll cover that here. We'll also look at how wine lists are actually written and explore questions to arm yourself with as you go into a restaurant or wine shop. This chapter also focuses on serving wine at home, including some service suggestions and pairing basics. There's a reason why storing your bottle of wine horizontally matters, why Sangiovese is a perfect match for your bowl of pasta all'Amatriciana, and why nothing could possibly taste better with your anchovies and pan con tomate than a glass of Galician Albariño or Txakoli. The way we eat and the way we drink pair naturally most of the time (although not, as wine grids would have you believe, mechanically). If you are in the mood to tinker with wine pairings, I've included a few ideas that are reliably delicious and will spark inspiration no matter whom you are hosting.

The final chapter, our "Last Sip," shares thoughts about where to go next if you're looking to dive in deeper and is a prelude to the appendix (page 230), which includes some of my favorite wines from around the world, as well as some of my favorite importers into the United States.

YOU ARE THE ARBITER
OF DELICIOUS

"They've really begun the war," he said to himself.
"And all over a word in a dictionary, the ninnies!"
—Natalie Babbitt, *The Search for Delicious*

"Delicious" is personal. It cannot be boiled down into a flavor grid or list of attributes. It is not a list of fruits and flowers that can be found, with a little training or imagination, on a grid. It will not show up with a sign on your doorstep advertising "99 Points" or materialize on social media with a cool-looking label. Trying to define your palate based on a flavor grid, points, or someone's social media feed can result in living within a world on someone else's terms, and you deserve better than that. Delicious is a value system that *you* decide on and refine as your tastes evolve.

As a young sommelier, I thought of each wine in context of this flavor/structure profile, a clinical and sometimes oversimplified understanding of its identity. As a more seasoned adult, I've outgrown that way of thinking. I've been asked by countless wine surveys if I prefer "apples or pears," "lemons or tangerines," and "hazelnuts or almonds," as if these questions were able to unlock my palate and point me toward a lifetime of wine-drinking fulfillment and pleasure. Following that path reduces your journey to a treadmill, in which you're meant to stay within a limited framework that keeps you drinking similar bottles your whole life, disregarding nuance and context.

My intention for this book is to help you identify your own wine value system and inspire you with questions to ask, which I believe is the best way to fine-tune your compass for drinking well and creating experiences that you'll love. Wines made from the same grapes in the same region will share a similar flavor profile—but some are flat and unexciting, and others are so soulful you will feel it in your bones. The wines I put on my lists as a sommelier and that I'm happy to purchase at full retail price share a value system based on a commitment to quality and a pattern of decision-making that begins and ends with the details. If you have had wines like these, then you can relate! I hope this book will help you learn what these wines are for you.

COMMUNITY AND CONNECTION

I believe wine's highest purpose is to bring us together, connect us to one another and to different parts of the world, and help us experience different points of view. A special bottle is infinitely more special when it is shared with someone you love, or someone who is just as excited about it as you are. That bottle becomes meaningful when there is a memory surrounding it or stories shared while drinking it. It becomes a moment in time. Whether it's an expensive bottle or not is often irrelevant. While I've always been someone who has enjoyed knowing the "why" behind wine—and much of this book explores those kinds of questions—wine's most important quality is to expand our horizons and our perspectives, at the dinner table and beyond. The greatest asset you can possibly have and hold on to throughout your journey is a sense of wonder.

"WINE EXPERTS"

Allow me to note that I snicker internally when I meet someone who identifies as a "wine expert." The most brilliant wine professionals have studied far too much to retain an arrogant attitude toward wine, a topic that no one singular person can know everything about. Studying wine is a humbling—yet exhilarating!—experience because the world of wine is so vast and ever-changing. The more open-minded and curious you can be when you're starting out, the more adventuresome your journey will be. And if anyone you meet calls themselves a "wine expert" before positing something, please immediately dismiss whatever they are about to say.

A BRIEF
HISTORY LESSON

We can thank the Phoenicians (who were based in modern-day Lebanon) for introducing wine and viticulture to ancient Italy and Greece, and we can thank the Roman Empire's army for planting grapes throughout Europe. (They were not necessarily gourmands; they just needed to stay hydrated, and hadn't yet refined the system of aqueducts to bring potable water to the masses). We can also thank generations of (mostly French) monks and their quest for excellence through a methodical process of trial and error—plus, of course, politics, extreme weather events, colonization, and conquest in war, by which grapes ultimately came to define certain regions. And while American wine culture began with a "manifest destiny" approach to regions and grapes—planting any kind of grapes anywhere and hoping for the best—the country, along with the rest of the "New World," is now learning which grapes perform best in which regions, leading to extremely compelling wines.

The United States' relationship with fine wine is quite new compared with that of Europe and the Near East, in countries such as Armenia, Egypt, and Georgia. In fact, no one in America could successfully grow *Vitis vinifera* (the species of grapes that makes wines you know, love, and have heard of) until the early nineteenth century, just a few generations ago. By comparison, winemaking in Europe dates to the Neolithic Age (12,000 years ago), even if things didn't hit their stride until 6,500 years ago. The country of Georgia has a viticultural history dating back 8,000 years.

America's budding wine scene did not have much time to flourish before Prohibition (1920–1933) brought it to a halt, and shortly thereafter the country plunged into World War II (1941–1945). By the time wine on any level emerged as a notable beverage, Americans were drinking E. & J. Gallo's Thunderbird, first released in 1957—a smash hit of high-alcohol white port mixed with sugar and lemon juice concentrate. We can thank Robert Mondavi and his contemporaries for recognizing that something better had to exist, traveling to France, and aiming to upgrade America's wine culture by essentially trying to recreate Europe's.

VITIS VINIFERA (AND PHYLLOXERA)

Many species of grapevines exist but only one, *Vitis vinifera,* makes grapes that produce wines you have likely encountered. *Vitis vinifera* was native to the area surrounding the Caspian Sea (a giant saltwater lake that touches Iran, Russia, Georgia, and Armenia), before spreading to Greece and Sicily, then throughout Europe. Whether you fancy Chenin Blanc or Gewürztraminer, Gamay or Pinot Noir, these all belong to the *Vitis vinifera* family that originated thousands of years ago.

America produces thirty-four different native *Vitis* ("grape") species that create largely undrinkable wine, described as "foxy" in wine books. (I have never smelled a fox so I can't verify this; I can, however, tell you they smell feral.) The most notable quality of these native grapes is their ability to peacefully coexist and grow alongside phylloxera (*fil-LOX-or-ah*), a root louse native to North America's East Coast that eats and destroys *Vitis vinifera* vines.

In the 1860s, live vine cuttings from America made their way to Europe, a fact generally attributed to the invention of steam ships, which traversed the ocean faster than previous boats, enabling the root louse to survive the journey into Europe and proceed, little by little, to devastate an entire continent of unsuspecting vines. Devastation first hit Tavel, in southern France—a small region known by sommeliers who traffic in exam notecards as the one region in France that exclusively makes rosé. Then these tiny bugs flew north, catching rides on the wind and gorging themselves, wreaking havoc on roots from the Rhône to the Rhine. France's vineyards fell first, followed by those of Italy, Germany, Austria, and eventually Spain. French winemakers Gaston Bazille and Leo Laliman independently proposed an idea that no one particularly liked: grafting French vines onto American rootstock. A Texas horticulturist named Thomas Volney Munson recommended grafting French vines onto Texas's own *Vitis berlandieri* rootstock, which grew in limestone soils similar to those in France. Grafting worked, and that was that. Today most of the world's vines are grafted onto phylloxera-resistant American rootstock in this manner. That discovery also allowed America to at last grow wine grapes and begin to make world-class wine. That was just over a century ago, a blink in the long history of humans making and drinking wine.

CHAPTER 1

TASTING (AND WHY WE'RE STARTING HERE)

A love of wine usually starts with a sip from a glass, so that's where we'll begin, too. The revelatory wine moments in my life all involved tasting a bottle that moved me to my core. Perhaps you're reading this book because you, too, tried a glorious glass that made you pause, reflect, and aim to recreate the experience. By starting here, I hope to arm you with a set of tools to unpack what's in that bottle you love—and help clarify why you enjoy it so much.

This chapter also happens to be the most technical in the book—it outlines the process by which some professional sommeliers learn to analytically taste. If it feels like too much too early, skip ahead, but know that it's here if and when you're ready.

WHY ANALYZING WINE IS ACTUALLY USEFUL

I've been on a Deepak Chopra kick lately, and I think the best way to explain why and how tasting wine analytically is different from just drinking a glass of wine is that it involves *attention* and *intention*. As with anything, the quality of *attention* and the underlying *intention* you give to something is directly tied to the depth of your relationship with it. Drinking wine for pleasure is done with the intention of enjoyment. You can stop there if you'd like! But if you'd like to know why you like it, that's where analyzing comes into play. Tasting wine analytically is all about understanding its composition. How does a bottle of Pinot Noir compare with others you've tried? When you focus on this, you'll start to notice things about the grape, region, climate, vintage, and production methods that you would never have thought to notice before. Tasting analytically involves paying attention to these different factors, and then exploring them further to understand the wine on a deeper level. Is this something every wine lover needs to do? Certainly not! Is it something that I believe will give you a lot of knowledge and confidence as you continue on your journey with wine? Absolutely.

There are many different approaches to tasting wine analytically, and they all bring you back to the ultimate goal: knowing how to max out your drinking pleasure in any scenario. Your palate is a muscle. The more you exercise it, the better you'll get at recognizing quality and understanding what you like in a wine and why. Like running, tasting is one of those activities in which effort is directly proportional to results. Anyone can get good with consistent practice. And the practice is fun!

Blind tasting is also, at its core, a game of logic, adding an element of probability to the often-mysterious world of wine. A great bottle is a time capsule—a liquid snapshot of the vintage, soil, climate, and even the winemaker's philosophy and point of view that year. Once you understand a few basics like a general profile of some key grapes (covered on pages 109–125), the way climate and soil impact a wine (see pages 92 and 97), and why farming matters (see page 94), you're well on your way to understanding what makes a wine express itself as it does.

One world-famous winemaker friend in Chambolle-Musigny, a village in Burgundy renowned for red wines made from Pinot Noir, shared that when he blind tastes wine, he sees shapes and colors. He has Rothko-like images in his mind for each wine he tastes! Another winemaker friend in Meursault (a village in Burgundy famous for white wines made from Chardonnay) told me he thinks analyzing a wine's aromatics is a waste of time; he prefers to focus exclusively on a wine's structure (tannins, acidity, alcohol—we'll get into that shortly, on pages 35 and 40). Another friend, a master sommelier, shared that she envisions types of flooring when she tastes. This one makes no sense to me, but for her, some wines are "hardwood floors" while others are parquet or tile. This is to say there is no one way to interpret wine in your mind's eye. The more you practice, the better you'll get at knowing what resonates with you.

HOW I ANALYZE WINE

Included below is the process by which I learned to analyze wine through blind tasting, in case you'd like to give it a try. "Blind tasting" is more than a party trick, and it's something that you—and anyone—can do well with practice. Tasting something "blind" just means you give a shot at guessing what it is without seeing the label. Handily, this practice removes any preconceptions, allowing you to actually taste the wine on its own merit instead of predetermining your opinion based on what score it has or whether your enthusiastic friend says it's great.

As with so many things in life, process is at least as important as the end result. The method below is by no means the only way to taste. But it's one way that works. And it covers the bases, giving you the tools you need to understand wine by really paying attention to its different structural components. You're going to hone your focus on one part of its identity at a time, beginning with the appearance, then the smell and flavor, and finally the wine's structure, arguably the most important of all. That said, you know how you learn best, and if this is too much too early, come back to it another time, or take what works and leave the rest.

The most notable qualities of a wine come from the grape itself, a topic so large it gets its own chapter (see page 105). Each grape has its own structural snapshot. This section on tasting is devoted to helping you hone your skills to recognize these nuances in sight, smell, and taste.

APPEARANCE

There is so much you can learn about a wine just by looking at it. I like to start with a flat sheet of white paper and a clean stemmed wineglass. (If you're giving this a try, sniff your wineglass—some look clean but smell like dishwater or dusty cabinets.) Your blank sheet of paper will work as a backdrop for evaluating the wine's appearance. Just place the paper on the table, countertop, or whatever surface you're using. Pour one or two ounces of wine into the glass. Raise it above the white paper, then tilt it to about 45 degrees so that the wine reaches out toward the rim. With the glass still slanted at 45 degrees, turn the stem in your fingers so that the wine stays in the same place but the glass moves around it, coating the sides in wine. While doing this, focus on three things: color, clarity, and viscosity.

COLOR

Color gives you some important clues about wine. It indicates possible grape varieties, a wine's age, and the vessels in which the wine spent time before bottling. The more you pay attention to different shades of greens, yellows, and golds in white wines—or reds, oranges, purples, and browns in red wines—the more you'll see the nuances. White wines usually range from greenish-yellow (young and vibrant, without time in oak) to brownish-gold (older and evolved, likely aged in oak barrels). New oak gives wine a golden color and some additional texture, thanks to the evaporation that occurs through the oak's pores. Older oak contributes a round texture but doesn't influence color.

Red wines range from purply-fuchsia (young and vibrant) to garnet-brown (older and evolved, and often aged in oak). That said, some grapes—like Malbec, Carmenère, and Syrah—naturally have purplish tones, even when they've been in the bottle for a decade. And other grapes—like Nebbiolo and Grenache—oxidize quickly, so their color has brown and orange hues, even when they're young. If this sounds strange, think of the way some sliced apples—like Granny Smith—stay whitish in color even after they've been sliced for a few hours, while other kinds of apples, like Red Delicious, start to turn brown almost right away. It takes a bit of practice to notice these distinctions, but you are the Sherlock Holmes of your tasting experience. When you start to look for these details, you'll begin to notice a world of new information just by glancing into your glass!

PINOT PINK

Pinot Grigio (aka Pinot Gris), a pink-skinned grape, lends a slight pink hue to the wine, a fun clue when blind tasting.

CLARITY

When analyzing clarity, you're getting insight into a few things, like if your wine is filtered or unfiltered, fined or unfined, and whether it's young or a few years old. In red wines, clarity can indicate grape variety (but this isn't the case for whites).

To analyze clarity, tilt your glass 45 degrees over a piece of white paper and look at the wine's reflection. Is it sharp and clear, like a shimmering star? Or is it muted? Wines casting a sharper reflection were probably fined or filtered. Cloudy wines with muted reflections were probably not. It's also a good idea to look for signs of sediment, which can be chunky, like crushed leaves, or dusty, like silt in a stream. Silty sediment can be a clue about age: older wines have some sediment, while younger wines do not.

While you're here, look for fine, granular sediment suspended in the wine. If you see any, the wine is probably a few years old. As red wine ages, color begins to precipitate (separate), and tannins polymerize (link together like a chain-link fence). This miracle of time creates solid particles, which are cute and tiny at first until they keep linking together, getting bigger and clunkier until they fall to the bottom of the bottle in chunks.

In red wines, thinner-skinned grapes like Gamay, Nebbiolo, and Pinot Noir are less pigmented and usually easy to see through. Other grapes, like Cabernet Sauvignon, Merlot, Malbec, Shiraz, and Cabernet Franc, produce opaque wines. One way to test for clarity is to write your name or draw a smiley face with a dark-colored pen on your piece of white paper. Now, pick up your glass of red wine and tilt it 45 degrees. See if you can read your name or your doodle through the most concentrated part of the wine in your glass (the thickest part, near the base). If you can see your scribbles, your wine was probably made with a thin-skinned grape. If you can't, the grape is more likely than not a thicker-skinned variety.

VISCOSITY

Viscosity is a term used to describe a wine's weight and texture. High viscosity manifests as stickiness on the sides of the glass and a rich and luxurious mouthfeel. Higher-viscosity wines leave more wine clinging to the sides, creating thicker, slower-falling tears. To analyze viscosity, gently spin the stem of your tilted glass one or two more times, then hold the glass upright and watch the tears fall. Tears in a rush to get back down indicate that the wine has a lower viscosity and a lighter body; wine that meanders has a higher viscosity and is fuller in body with more alcohol.

Contrary to popular belief, the tears' thickness has nothing to do with quality and everything to do with the amount of alcohol (or, in sweet wines, sugar) in the wine. Alcohol and sugar amp up viscosity, so if you have a wine that is clinging to the sides of the glass, it's probably high in alcohol. The grapes were probably grown in a warm climate or a moderate climate in a really sunny year. Wines higher in viscosity are fuller in body and richer in style. The opposite is true for low-viscosity wines.

SMELL AND TASTE

Without smell, there is no taste or flavor. The smell and taste department is where people can get carried away, waxing poetic about "new tennis ball smell" and whatnot. While you're welcome to stretch your creative limits here, the truly applicable portion of this exercise is just honing your focus and attention. Think of it like meditation, but with wine. When you really spend time with a glass and begin to notice the kinds of smells that remind you of fruits, flowers, herbs, and spices, you start building a database that will help you understand your very own aromatic profile for Pinot Noir versus Cabernet Franc, and then more specifically, Pinot Noir from France as opposed to Pinot Noir from New Zealand or California or Oregon or wherever. And before long, you'll be able to determine which Pinot Noirs are Premier or Grand Cru within specific villages in Burgundy. Most importantly, though, you'll get to know your own palate.

A mantra of my life during exam preparation became "fruits-earth-wood-flowers-herbs-spices-other," a mental checklist and reminder to search for each of these descriptors whenever I sat down to taste a glass. If you focus all of your attention on almost any wine, you'll identify things from each of those categories, and together, those descriptors will paint the picture of what you're drinking. While the mantra addresses the general categories, getting granular is the way to really sharpen your skill. For example, imagine you smell cherries in a glass of red wine. Qualify that. Go the next step and ask yourself what kind of cherries. Are they red, tart cherries? Or are they juicy, black Bing cherries? Are they just barely ripe? Or ripe and candied, like cherry jam? Ask yourself how these aromas are showing up in your glass: dried, baked, fresh, ripe, underripe, jammy, and so on. This will help you paint the picture in your mind's eye. Just barely ripe fruit indicates that the grapes were grown in a cooler growing region, whereas ripe, jammy fruit usually indicates a warmer one.

When you're ready to give this a try, lift your glass to your nose and smell the wine in a few different parts of the glass (ideally in one big inhale) to capture all of those initial aromas in your memory. After you've taken a mental (or written) note, go back and smell your glass again. It may sound strange, but you'll get different aromas from different parts of the glass. One important thing to remember: The first time you inhale, you're going to have the strongest perception and clearest picture, because that's how our brains work! Some of those initial aromas you won't be able to smell again for several seconds.

FRUIT

In white wines, the following are often present: orchard fruits (apples, pears, quince), stone fruits (peaches, plums, cherries, apricots, nectarines), citrus fruits (lemons, limes, oranges, grapefruit, mandarins, tangerines, yuzu), and tropical fruits (kiwifruit, pineapples, bananas, mangos, lychees). In red wines, I usually smell some combination of berries (strawberries, blueberries, blackberries, raspberries, red and black currants) and stone fruits (cherries, red and black plums, apricots, peaches). This list is by no means exhaustive! And, as mentioned above, the next step is to qualify these fruits. Are they tart and unripe? Just barely ripe? Overripe and juicy? Baked and almost jammy? These details will help you understand not only the type of grape but also the climate in which it was grown.

EARTH

The term *earth* covers both organic descriptors (roots, leaves, damp soil) and inorganic ones (rocks, stones, pebbles). The "earth" bucket is a reminder to look for a presence of rocks and minerals as well as earthy tones like roots, leaves, tobacco, and even *Brettanomyces*, a common yeast that is technically considered a wine flaw but that has captured hearts with its aromas of "barnyard" and forest floor.

The more you taste, the more you will be able to identify how grapes grown on different types of soil show up in your glass. I didn't notice this immediately. Admittedly it took me years to even think to pay attention to things like where I physically tasted tannins and acidity. But once I did, I began to notice that different grapes behave in different ways on my palate, and those same grapes grown on different soils present themselves with nuance, too. For me, there is a lift and perception of acidity that I notice on the sides of my tongue when grapes are grown on decomposing granite (e.g., many wines grown in the Northern Rhône Valley). When I taste wine from grapes grown on limestone, I perceive acidity nearer the front as well as the sides of my tongue. While you certainly don't have to geek out in this way, you can—and it can be quite fun! I'm sharing it here to spotlight the ways in which so many factors, some obvious and some not at all, play a role in that final wine in your glass.

OAK

Oak brings another dimension of texture and depth to a wine. It also affects flavor in interesting ways. American and French are the two most common kinds of oak used in winemaking. (We'll cover this in more detail on page 79.) American oak smells like vanilla and, oddly enough to me, dill pickles. French oak is a different species entirely and often reminds me of caramel and vanilla. It's also harvested differently; French oak has to be split with an axe, whereas American oak can be hewn with a saw, adding rougher edges and, arguably, more bombastic aromas and flavors. After oak staves are shaped into barrels, the barrels are toasted, and the level of toast (light, medium, or dark) influences a wine's flavor, too.

FLOWERS

Here is your chance to do a mental deep dive through your own imaginary flower shop. It's also your opportunity to learn about the world of monoterpenes (see page 34), a group of organic compounds that give smells, well, their smell. Whether you're getting scents of apple blossoms or perfumy gardenia or savory sniffs of sunflowers or squash blossoms, you can thank this particular class of hydrocarbons for that. Lucky for you, there are subcategories of monoterpenes that produce lots of familiar smells, opening up your world of wine to as many scents as there are blooms in a Parisian flower shop.

HERBS

In your wine-smelling journey, you'll probably encounter wafts of sweet herbs (mint, basil, ginger, chamomile) in some wines, and savory ones (parsley, sage, rosemary, thyme) in others. Of course, you are not actually smelling these herbs, you're smelling compounds like terpenes and pyrazines that give some wines the aromas of herbs, flowers, and spices.

SPICES

Spices like black peppercorns tend to show up in Northern Rhône Syrah for me, while white peppercorns are a staple in many Austrian Grüner Veltliners. Spices are also one more reminder to look for warm, toasty baking spices like cinnamon, nutmeg, vanilla, and clove, which are all indicators that the wine has spent time in new oak.

One side benefit to the exercise of analyzing aromas is that it may very well help you choose the perfect wine for dinner! How convenient, for example, that Austrian Grüner Veltliner can tend to smell of white peppercorns and green beans, or that Sangiovese tends to smell of oregano and dried tomatoes.

OTHER

Here is where you scan for smells and tastes that don't fit into any of the other categories. I put lees and umami-like things in this bucket. Lees are just dead yeast cells that can add a smell of white bread, stale beer, or, in the best cases, brioche. They also add texture—lees are an inexpensive alternative to oak for adding weight to a wine. A strong lees-y smell indicates the wine was probably made in stainless steel, and the winemaker didn't really have other resources like oak (which is expensive, around $1,250 per 225-liter barrel!). "Other" is also the place where I'd look for anything funky that may indicate the wine is off in some way.

Terpenes, Pyrazines, and Molecules That Give Smells Their Smell

If you are at the point in your wine journey where you want to take the next step and really level up, you are ready to learn about the world of molecular compounds responsible for that menagerie of flavors in your glass. Terpenes and pyrazines are fragrant hydrocarbons, which are organic compounds consisting of hydrogen and carbon. Knowing about them can help shine a light on why certain wines taste like a bridal bouquet while others taste like gazpacho and still others like a pastrami sandwich on rye.

Monoterpenes (mono-TUR-peens)

A subcategory of terpenes, monoterpenes are a class of chemical compounds that creates extremely floral aromas in wine. Gewürztraminer, Torrontés (primarily from Argentina), Riesling, Muscat, and to a lesser extent Albariño all have terpenes in their DNA. At their most pronounced, monoterpenes smell like a perfume shop. In subtler quantities, they can smell like orange blossoms and tangerines. Linalool is a rose-smelling terpene that pops up in Muscat, Riesling, and Albariño. Geraniol is another terpene that surfaces in Gewürztraminer, making it smell like a Diptyque "Tubéreuse" candle.

Rotundone (RAH-tun-dun-eh)

This molecule belongs to a different subcategory of terpenes called *sesquiterpenes* and shows up in black and white pepper, and, to a lesser extent, marjoram, rosemary, basil, and thyme. Sticking your nose into a glass of Northern Rhône Syrah will make you swear you're within arm's length of a pastrami sandwich, and you can thank rotundone for that. That same molecular compound is also responsible for the distinct white-pepper aroma in Grüner Veltliner.

Pyrazines (PEER-uh-zeens)

These are a different class of molecular compounds that smell of green bell peppers, fresh-cut grass, and sometimes green beans. Pyrazines run rampant in Cabernet Franc, a red grape prolific in the Loire Valley, Bordeaux, and throughout the globe. Cabernet Franc is also a parent of Merlot, Cabernet Sauvignon, Malbec, and Carmenère, red grape varieties that all have pyrazines present in varying capacities. Sauvignon Blanc (the other parent of Cabernet Sauvignon) is also chock-full of them—which is one reason it tastes so grassy.

STRUCTURE

Many sommeliers and winemakers believe structure is the most important part when tasting a wine. My mental-checklist acronym for structure during my exam prep was TAAL, reminding me to assess Tannins, Acid, Alcohol (including body), and Length (or complexity). While the smell and flavor of a wine are two sides of the same coin (try tasting something when you can't smell it), structure is a different thing entirely. The structure is how a wine feels on your palate. It's also typically the place where all the other clues you've been assembling about the wine coalesce.

TANNINS

Found primarily within the bark, fruit, and leaves of woody flowering plants, tannins are naturally occurring molecules known as polyphenols. In wine, tannins come from the grape skins and stems (and sometimes oak barrels) and are responsible for the drying sensation on your tongue, similar to the feeling you get when sipping oversteeped tea. Tannins are generally not present in white wines or rosés because the grape pulp doesn't spend much time in contact with the skins or stems. White wines aged in new oak barrels have a little tannin, which shows up as a subtle grippy feeling on the back of your throat as you swallow the sip. Orange wines (those made from white grapes but fermented like red wines; see page 53) do have tannins because the grape pulp spends days or even weeks macerating with the grape skins and stems.

Different red wines have different amounts of tannins. The main determining factor here is the grape. For example, Cabernet Sauvignon and Nebbiolo have high amounts of tannins. Other grapes, like Gamay (the grape of Beaujolais), have tannins so soft and gentle you will want to lay your head on them at night.

An admittedly weird-sounding but very effective trick for identifying tannins is to trap some wine between your top row of front teeth and your gums and let it hang out for a few seconds before you drink or spit it out. High-tannin wine will grip your poor gums like barnacles. Low-tannin wines will not. This odd technique helps identify the gradient of tannins from high to medium to low, and it works every time.

Wine Flaws

Wine flaws are worth the time investment to understand so that you'll know if something is wonky with your wine before you drink it! No one wants to be the person lyricizing about a corked bottle while others at the table whisper under their breath. That said, you also don't want to be a wino-chondriac, a person who learns about a wine flaw and then ascribes it to every bottle they taste. So proceed through this section with more curiosity than caution. And if these wine flaws are new to you, you can start to develop your nose for these by asking your favorite wine shop or restaurant to save you any bottles that are flawed so you can smell for yourself and learn from them.

Cork Taint (TCA or 2,4,6-Trichloroanisole)

Smells Like	Wet, moldy cardboard, or a wet dog. It can also show up as an absence of fruit with a spiked and imbalanced acidity.
Caused By	A few things. Historically, compounds naturally present in the cork bark (used to make corks) came into contact with fungicides and insecticides, which, unfortunately, were widely used from the 1950s through the '80s. Even though they're rarely used now, the resulting compounds remain in the soil and can still contaminate the cork—one more good reason to stop using chemicals! TCA can also occur when bleach reacts with lignin, a naturally occurring compound in cardboard or wood. Mold, yeast, and bacteria then convert this compound into TCA. (Once you've trained your nose to identify cork taint, you're going to smell it everywhere—from "corked" apples to broccoli—thanks to their transport in "corked" cardboard boxes! Sorry in advance—corked apples and broccoli are super annoying.)
How to Prevent It	Train your nose to identify this "corked" smell and send back wines that smell like rotten cardboard or a wet dog. You could also avoid wines sealed with cork, but I would never recommend that, as too many (in fact, all) of the greatest wines I've ever tasted have cork closures! Wines closed with screw caps open up the possibility for mercaptans (see page 39), so no option is perfect.

Oxidation

Smells and Looks Like	Prematurely dried fruits in a young wine, vinegary notes; brownish-colored wine.
Caused By	Too much oxygen at the wrong time, usually a result of a faulty cork or heat exposure.
How to Prevent It	Store wines away from heat and ideally somewhere humid if it's a special bottle you're planning to keep around for a while. Store the wine horizontally so the cork is in constant contact with the wine and doesn't dry out! Also, make sure you purchase bottles from a place that has ensured they've been stored properly.

Maderization

Smells and Looks Like	Toasted walnuts, baked fruit, raisins, and prunes, sometimes with a hint of vinegar. The color is usually brownish.
Caused By	Exposure to heat (e.g., left in a hot car or stored in a kitchen cabinet).
How to Prevent It	Keep your wine out of cabinets and garages! And purchase any fancy bottles from a reputable source.

37

Lightstrike

Smells Like	Old cabbage, wet newspaper, and light sewage.
Caused By	Blue and ultraviolet rays transform amino acids into sulfides (not to be confused with sulfites) and other smelly compounds. (*Sulfides* [S^{2-}] are volatile sulfur compounds that smell of rotten eggs. *Sulfites* [SO_3^{2-}] are oxidized sulfur-containing compounds.)
How to Prevent It	Don't buy wine in clear glass bottles, especially from supermarket shelves or places likely to have blue or ultraviolet rays lighting the aisles!

Mousiness, or Goût de Souris

Smells Like	Author Alice Feiring describes this as "puppy breath" and "dog halitosis"—it's hard to improve upon that description.
Caused By	A lactic bacteria taint that can be hard to smell but becomes very noticeable on the palate. It's typically caused by low use of sulfur dioxide.
How to Prevent It	If you are buying a natural wine, know this is a risk unless you're already familiar with the winery.

Volatile Acidity

Smells Like	Vinegar or nail polish remover.
Caused By	Often called VA (short for volatile acidity), this flaw refers to the vinegary smell resulting from too much acetic acid, a type of acidity that occurs naturally during fermentation in small quantities but gets overwhelming quickly. In the European Union, it is not legally allowed at quantities more than 1.2 g/l for red wines (or 1.08 g/l for rosés or whites). The acid itself isn't the problem; rather, the smelly smells occur when a strain of bacteria (*Acetobacter*) feeds off oxygen and creates a reaction with alcohol in the wine.
How to Prevent It	This is the winery's job! That said, VA is not always a bad thing, and many people like a small amount in a wine. It tends to be somewhat common in Italian red wines, particularly from warm regions like Montalcino or Taurasi.

Brettanomyces

Smells Like   Barnyard or horse manure.

Caused By   *Brettanomyces,* or Brett as it's often called, is a yeast strain often associated with the smell of cows, leather, and animal poop. Previously common in warmer growing regions, it's common everywhere now, and is usually found in red wines. Some people love Brett, while others view it as a sign of bad winery hygiene.

How to Prevent It   Wineries that keep good hygiene and use sulfur dioxide tend to avoid an overload of Brett. Cold wine regions and underground cellars also help keep strong Brett aromas at bay.

Reduction, Especially Mercaptans (or Thiols)

Smells Like   Sewage, rotten egg, old cabbage.

Caused By   Winemaking with too little oxygen exposure. There are many volatile compounds produced during winemaking, but one of the most notably problematic ones is hydrogen sulfide, caused by yeast reacting with sulfur after fermentation.

How to Prevent It   The winery is tasked with this and can usually avoid mercaptans or thiols forming by racking and aggressively adding oxygen to the wine. This particular flaw shows up most often in screw-top bottles, which inhibit oxygen flow.

ACID

A good way to think about acidity is how much the sides of your tongue curl and your mouth puckers when you take a sip. Some wines, like a dry Riesling from the Mosel region in Germany, have acidity so high you might as well be eating a lemon. Others, like Gewürztraminer, have almost no acidity at all. For context, Chardonnay falls in the middle. Another thing about acidity: There are several different kinds, and they all taste different. Malic acid is sharp, like a Granny Smith apple. Lactic acid is softer, like plain yogurt or lemon curd. Acidity is dynamic and different for every glass of wine, let alone every grape!

ALCOHOL

Alcohol gives wine a lot of its weight, body, and texture. It is also a conduit for flavor, sort of like fat is for food. Alcohol is a direct result of how much sugar was in the grapes before fermentation, since yeast eats sugar, converting it to alcohol during fermentation. Hot regions, like Priorat in Spain or the Barossa Valley in Australia, tend to grow ripe, juicy grapes that turn into wines with a high alcohol content. Cool regions like Champagne, France; Santa Barbara, California; and Tasmania, Australia, grow grapes with less sugar—and make lower-alcohol, lighter-bodied final wines.

Assessing alcohol helps to indicate whether a wine was grown in a warm region or a cool one, a sunny season or a rainy one. To do this, smell the wine intensely—with your eyes open!—as close to the glass as your face will allow. Don't worry, no one is watching you. A high-alcohol wine will nearly burn your retinas if your eyes are close enough to the glass.

LENGTH AND COMPLEXITY

Two final aspects to consider when analyzing a wine are length and complexity. Length refers to the amount of time the wine's flavor lingers on your palate. Complexity asks you to consider the layers and nuance in a sip. Some wines, like the ones in jugs at the bottom of supermarket shelves, have a flat, acrid aftertaste, as though you've drunk water steeped with decomposing wildflowers. Others might also taste oddly sour, as though someone dissolved an aspirin tablet in it.

Not to sound judgy, but wines that wither on your tongue were likely made from overcropped, high-yield grapes harvested by machine from chemically treated soil, then adjusted with more chemicals in the "winery" (which is probably far more industrial than you'd imagine). These wines are not aging in lovingly crafted oak barrels or cement vats, and any tannins or acidity in the wine were added as a prepurchased powder. By contrast, a well-made bottle will offer a lot of complexity—its texture and flavors lingering on your palate for quite a long time.

The more you taste, the more you'll be able to see, smell, and taste things that you couldn't before. Best of all, you can decide when it's time to analyze or time to enjoy whatever is in your glass. Of course, they're not mutually exclusive!

PUTTING IT ALL TOGETHER

Repetition creates confidence, in analytical tasting as in anything. Use your new skills to get a clear picture of any wine you taste. The more you try, the more reference points you'll develop, and the more opinionated you'll become about which wines bring you joy. Now that you know what to look for, you have an arsenal of tools at your disposal to decipher clues that link together and create a picture of a "typical" wine from any given classic region. Soon, you'll be a pro at piecing together the individual puzzle pieces and understanding the big picture in each wine—and refining your palate and preferences every sip of the way.

DEFINING DRY

One of the more confusing wine terms is *dry*. Any wine allowed to finish fermentation is dry, meaning there is no unfermented sugar left. In Europe, the EU Commission Regulation defines dry wines as those containing no more than 9 grams per liter of residual sugar (except in cases where acidity is over 7 grams per liter as well).

Dry just means the absence of sweetness. Dryness is not an anomaly, nor is it any indication of quality.

Off-dry means a little bit sweet and is the perfect term to use in a wine shop if you're looking for something with a hint of sweetness—say, to pair with a spicy meal—but you don't want a full-on dessert wine.

Sweet wines are high in sugar, and most of the time they are also higher in sulfur dioxide, to stop any potential refermentation. If you are sensitive to sulfur dioxide, note that sweet wines tend to have a bit more than other styles.

CHAPTER 2

WINE STYLES FOR YOUR DRINKING PLEASURE

Did you know that wine doesn't have to come from grapes? Technically it can be made with other fruit or even plants as obscure as dandelions (yes, dandelion wine is a thing—google it!). But serious wine—the kind in any bottle labeled simply *wine*—is made from fermented grape juice. Grapes probably evolved as the ingredient of choice because they contain the highest sugar content of any fruit except dates, which are notably short on juice.

This chapter presents a general overview of the different types of grape wines. We'll spend time on how sparkling, white, red, rosé, orange, and yellow wines are made and what to expect from them. My hope is that this chapter also inspires you to expand your horizons with new liquid adventures like sweet wines (yes, really!) and even fortified options. We'll also talk natural wine here. (You can find my thoughts on natural wine on page 45.)

HOW FERMENTATION WORKS

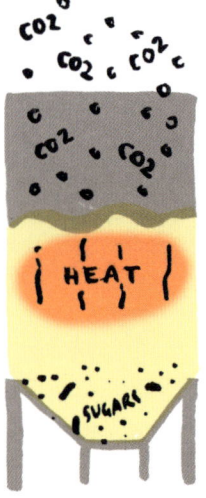
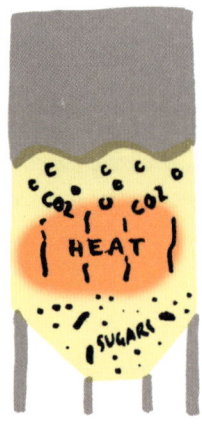

Yeasts, which are microscopic single-celled organisms that live in wineries, on grape skins, under the ocean, on your skin, and pretty much anywhere that isn't a sterile surface, are responsible for turning grape juice into wine. The kind of yeasts particularly attuned to this task are called *Saccharomyces cerevisiae*, and they love to eat sugar. They devour it in grape juice, metabolizing the naturally occurring sugar present in grapes into alcohol. The sweeter the grapes are at harvest, the higher the alcohol content in the final wine. As yeasts metabolize sugars, they release heat and carbon dioxide. In still wines, the carbon dioxide escapes. In sparkling wines, it's trapped in the bottle, resulting in bubbles.

STILL **SPARKLING**

SUGAR

A lot of friends ask me which wine they should drink if they're trying to avoid sugar. This sidebar is meant to help clarify the difference between drinking a glass of wine and, say, eating a bag of Skittles. For one thing, if you are drinking a dry wine (which is any wine aside from a dessert or sweet wine), you are not even the one consuming the sugars! The yeasts have already done that for you. For another, you are consuming a beverage made from fermented naturally occurring sugars in grapes, no chemicals or laboratories needed. *Sugar* is a charged word—and to be clear, processed sugars are something I largely avoid! That said, a bag of candy made with refined sugars and then colored and flavored with artificial additives is hardly analogous to drinking a glass of good wine.

The amount of alcohol in a wine is directly related to the amount of sugar in the grapes at harvest. The more sugar present in the grapes, the more alcohol there will be after fermentation. Sunny, hot regions produce grapes with more sugar because more sunlight leads to a higher sugar content, thanks to photosynthesis (see page 66). You can expect a bottle of Shiraz from the Barossa Valley in Australia to have a higher alcohol percentage than a Syrah (same grape—the Australians just call it Shiraz) from Côte-Rôtie in the Northern Rhône, which is cooler and less sunny.

Fermentation stops when yeasts can no longer eat sugars. Sometimes this happens because the yeasts have consumed all of the sugars and run out of food. (In this case, they starve, die, and fall to the bottom, where they're better known as *lees,* adding texture and a brioche-y quality to the wines.) Other times, the yeasts convert so much sugar to alcohol that they can no longer survive in the high-alcohol environment. Sometimes a winemaker will intervene to stop fermentation either through sterile filtering the wine (pressing it through a tiny microfilter that doesn't allow the yeasts to pass), or by adding something to kill the yeasts outright, like Velcorin or sulfur dioxide.

NATURAL

There's a lot to love about the roots of the natural wine movement. In theory, natural wine is a rejection of hidden chemicals and lazy overuse of sulfur as a preservative, a celebration of organic and biodynamic farming, and a counterculture attitude against wines made in factories with additives at every turn. In practice, *natural* is an increasingly polarizing and ultimately meaningless word in both the food and wine worlds. A lot of the fiery dogma behind this movement is accompanied by a lack of actual winemaking experience, and while I have sought out and purchased some excellent natural wines for my own cellar and for wine lists I've written through the years, I've also been taken aback by the lack of knowledge and general combativeness that natural wine, as a movement and a mood (at least in the States), has come to represent.

Even winemakers who once embraced and helped define the natural wine movement now shy away from the term as it's become so charged. Sicilian winemaker Arianna Occhipinti, an early proponent of natural winemaking, now distances herself from it in interviews and says she strives to make "good wine" instead. The author and journalist Alice Feiring, who largely brought awareness of the natural wine movement to the United States, catalogs with bewilderment how flawed wines plagued with bacterial infections and souris (see page 38) now flood natural wine tastings, particularly in the States.

If you are like most people who care about your body and the planet, you might assume that natural wine is not only a *good* thing to drink but that it is the *right* wine to drink, because "natural" is a benevolent-sounding word and implies anything else is somehow "unnatural" or bad. However, the term *natural wine* was born just a few decades ago, and it's not regulated anywhere except France, which happened only as recently as 2020. Natural winemakers don't have a legal obligation to work with organically farmed grapes or even to keep additives out of their wines. Of course, the highest-integrity natural winemakers do make those choices, but they're not mandatory, and no one is checking. Moreover, wineries that have been making wine without chemical additives for more than a decade or two would never attach themselves to such trendy terminology. Domaine Jean-Louis Chave in the Rhône Valley has been making wine without additives since 1481 and springs to mind as one example of a winery that predates the term and doesn't identify with it, yet emblemizes everything the movement strives to represent.

I'm troubled by some "natural" wineries in the United States that work with conventionally farmed grapes (meaning they use chemicals and don't tell you—see page 94) and that quietly dose their wines with Velcorin, a neurotoxin preservative that must be administered with a hazmat suit and breaks down to CO_2 and methanol, a highly toxic alcohol (that can be liquid or gas) even in small quantities. Velcorin flies under the radar because it was registered as a "yeast inhibitor" through a US Food and Drug Administration (FDA) loophole and therefore does not have to be disclosed on labels in the United States. (It's also prevalent in hard seltzers, energy drinks, juices, and many other things that are not naturally shelf-stable. As an aside, we make RAMONA in Italy instead of the States so that we don't have to use Velcorin, which seems to have an iron grip on the American beverage industry!)

To make things even more confusing, the word "natural" cannot actually be listed on bottles anywhere in the European Union. Instead, look for French wines carrying the designation "Vin Méthode Nature." Wines labeled in this way must meet the following criteria:

1.   Hand-harvested certified organic grapes

2.   Fermentation through naturally occurring yeasts that are present in their vineyard or winery (as opposed to purchased yeast grown in a laboratory)

3.   No additions (such as tannins, acid, sugar, water, color, or any other additives)

4.   No manipulations (such as reverse osmosis, which can be used to reduce the amount of alcohol or filter out any other unwanted aspects of the wine)

5.   No added sulfites before or during fermentation (sulfites can be added in the finished wine within the limit of a maxium 30 mg)

WHITE

White wines are made by fermenting freshly pressed grape juice. This juice can come from any color grape, although as you may have guessed, usually white wine is made from yellowish-green-skinned grapes. While it's less common, white wine can also be made from pressed red grapes. Color and tannins in wine come from letting the grape skins macerate in (hang out with) the juice, so pressing red or pink (aka "gris" and "grigio") grapes before they lend color to the juice will result in a white wine as well.

RED

A critical difference between white and red wines is in the production. While white wine is made by fermenting pressed juice, red wine is made from fermenting grapes (skins, seeds, and all) and then pressing those grapes *after* fermentation. The longer the juice, grape skins, and pulp macerate (which can be over a period of days, weeks, or even months), the more color and tannins from those grapes end up in the final wine.

ROSÉ

Rosé, a recent global phenomenon in that it went from being a wine industry afterthought into a powerhouse category in just a few decades, currently accounts for 10.5 percent of global still wine consumption.

Rosé can be made in one of three ways:

1. Blending white and red wine together

2. Allowing red grapes to macerate in a tank briefly, adding a pink color to the juice before the grapes are pressed and fermentation continues as though making a white wine

3. Siphoning off a portion of pink juice from a tank of fermenting red grapes, then fermenting that separately—a process known as the *saignée* ("bleeding") method

The saignée technique has a twofer effect: it concentrates color and flavor in the future red wine by creating a higher grape-to-juice ratio, while also creating a rosé from the excess juice. A bonus for everyone! For the winery, this method serves the dual purpose of producing a more powerful, structured red wine from which the rosé "bled," as well as generating cash flow (and a refreshing summer beverage)

How wine is made ...
WHITE

BARREL

BLADDER

SETTLE
+ RACK
(Remove juice from solids)

HARVEST

CRUSH
and/or DESTEM
(OPTIONAL)

PRESS

PNEUMATIC

RED + ORANGE

HARVEST

CRUSH
and/or DESTEM
(OPTIONAL)

GRAPES   STEMS

Ferment
with skins

PRESS

ROSÉ

HARVEST

DESTEM
(OPTIONAL)

Ferment
for short time
with skins

Then process as
for white wine

PRESS

SETTLE

Saignée method

Bleed off some of
the juice to ferment
without skins or seeds
(red wine is made
with the remainder)

Ferment
without skins

FERMENT
CLEAR JUICE

Malic Acid
is converted
to Lactic Acid

AGE

RACK

FINE + FILTER
(OPTIONAL)

BOTTLE   CORK   LABEL

AGE

FILTER

RACK

FINE + FILTER
(OPTIONAL)

BOTTLE   CORK   LABEL

FERMENT
WITHOUT SKINS

AGE

RACK

FINE + FILTER
(OPTIONAL)

BOTTLE   CORK   LABEL

with a rosé wine that can go to market just a few months after harvest. Saignée rosés often benefit from investments like hand-harvested, perfectly mature grapes, as those grapes were usually initially destined to become a fancier red wine.

SPARKLING

Wines sparkle when they contain trapped carbon dioxide. While it took a few centuries to get where we are today, we're at last living in a golden era of bubbly beverages where these wines have reached a nearly universal beloved status. There are a few different ways to achieve bubbles, and each method produces a different style of sparkle at vastly different price points. A few of my favorites are listed below.

MÉTHODE ANCESTRALE

The oldest method for making sparkling wine is fittingly called méthode ancestrale ("ancestral method"). It involves bottling a wine midway through a single fermentation and letting that fermentation continue in the bottle until it can't anymore. This happens either because the yeasts run out of sugar to eat or the alcohol level gets too high for them to survive. The méthode ancestrale is relatively simple and imprecise. It produces lightly sparkling wines that can be sweet, bone-dry, or even a little funky. Pét-nat wines, short for *pétillant-naturel* (naturally a little bit sparkling), are made this way. The oldest-known sparkling wine is called Blanquette de Limoux, made in the méthode ancestrale from the Mauzac grape in Limoux, France. You can find them in stores today. They're fun to have on hand as a refreshing dessert or even an aperitif option when you want to mix things up.

CHAMPAGNE (AND THE TRADITIONAL METHOD)

The most famous and complex way to create bubbles is called the traditional method (also called the méthode traditionnelle, or méthode champenoise), a process required in Champagne production and common in other sparkling wines that want to emulate the style. *Champagne* is a protected term. The only wines allowed to use it on labels must be made from approved grape varieties grown specifically in the Champagne region of France. Producers have to follow a particular production process where the second fermentation occurs inside the final bottle. (The exception to this labeling rule is American wineries that co-opted the term prior to 2005 and were grandfathered into a trade agreement.)

In the traditional method, barely ripe grapes are harvested, pressed, and then fermented to make a tart, still base wine that my husband likens to battery acid (it's called the vin clair). Next comes the blending (or assemblage). Historically, Champagne as a region turned its challenges (a cool climate with fickle ripening patterns) into strengths—blending different grapes from different vineyards, regions, and vintages to create a "house style."

Once the blend is set, the wine gets transferred to a glass bottle with a dose of what's known as the liqueur de tirage—a liquid-y mix of yeasts, sugar, and clarifying agents. Yeasts eat these sugars, which creates that famous second fermentation known as the prise de mousse (foam creation) that continues for about eight weeks. The carbon dioxide released stays trapped inside the bottle, along with the dead yeast cells (aka lees, or *lie* in French).

One essential part of Champagne production involves letting the wine spend time in contact with the lees. This adds texture as well as the region's trademark brioche-like flavor. The wine must spend at least fifteen months in that bottle from the date of that second fermentation to be labeled Champagne, and a minimum of three years to be labeled Champagne with a vintage date. Then autolysis (*aw-TALL-uh-sis*) works its magic. *The Oxford Companion to Wine* defines autolysis as "the destruction of the internal structures of cells by its own enzymes." Essentially the dead yeast cells begin to eat themselves, and the chemical changes

A HISTORY OF BUBBLES AND STRONGER GLASS

Before humans understood fermentation, no one realized that yeasts simply go to sleep when temperatures drop. Winemakers just assumed fermentation had finished when winter arrived. When warm temperatures returned in the spring, yeasts woke up and reactivated, and fermentation continued—often to the detriment of many not-very-strong glass bottles! Exploding bottles were a major problem before the 1630s, when the British knight Sir Kenelm Digby invented stronger glass by smelting coal instead of wood for hotter flames.

that occur during this time are pretty amazing. Autolysis stabilizes the wine, upgrades its mouthfeel by making the bubbles tinier and the texture silkier, and even prevents oxidation, so the wine can essentially age sur lie ("on the lees") in perpetuity until disgorgement, when it is separated from the lees.

While sur lie, the bottles are stored neck-down in riddling racks (wooden boards propped against each other like an A, with circular holes cut on a bias to hold bottles pointed upside down). The bottles are turned briskly at a steeper angle each time (riddling) every couple of days until the neck is pointed to the ground and the lees have collected in the neck, thanks to gravity. Then the bottle is disgorged (plunged into an icy bath that freezes the lees into a pellet, which is expelled like a tiny cannonball when the bottle is opened). At this point, the bottle is topped with a wine or a diluted mix of sugar or filtered grape juice (dosage) to make up for what it lost during disgorgement. The dosage's job is to round out the texture, focus the flavors, and in some cases add a little sweetness. Bottles labeled "zero dosage" use Champagne to top up the bottles instead. (Zero dosage is trendy right now—don't let anyone tell you dosage is inherently bad! You have to do a little work to determine your own preference.) Then the wine bottle is corked, secured with a wire cage, labeled, and released into the world.

Other regions throughout France—and the globe—use this same technique to make high-quality sparkling wines. In France, any bottle labeled "Crémant" has been made in the traditional method but in a region that isn't Champagne. Crémant bottlings require less aging sur lie—nine months instead of twelve. Franciacorta, from Italy, must be made in the traditional method (and requires even more lees aging than Champagne, something the region seems to value more than consumers do). Certain ambitious Cavas from Spain, Sekts from Austria and Germany, and well-made sparkling wines from the United States follow Champagne's inspiration, too.

CHARMAT, OR TANK METHOD

Every other method of sparkling wine production is much simpler, by design. The Italians created the Charmat method (also known as metodo charmat, Martinotti method, or the tank method) to make wines like Prosecco and Moscato d'Asti, which emphasize fruity instead of brioche-y notes. This process is also much less time-consuming and expensive than the traditional method, and that discount gets passed along to you, the consumer. Hooray!

In the Charmat method, a still base wine is mixed with sugar in pressurized, enamel-lined tanks to create a bulk secondary fermentation. Generally, no minimum lees aging is required, although Moscato d'Asti requires at least one month aging in tank sur lie.

A geographic area and legally defined AOC, confusingly called Château-Chalon (sounds like a winery but it's a region, I promise), produces the most famous vins jaunes (yellow wines), and the wines are always bottled in a vessel called the clavelin, which is smaller and stockier than a regular wine bottle at 620ml (versus the usual 750ml). Supposedly, this reduced volume amounts to wine remaining after evaporation during the six years and three months in cask. As the buyer, you are essentially acquiring a curious-shaped bottle and an evaporation tax of 130ml.

CARBONATION

Lastly, some wines, such as the five-dollar bottles of American-made "Champagne" I drank in college, are just still wines injected with CO_2. This is the fastest and cheapest way to get bubbles.

YELLOW

Vin jaune, or yellow wine, is the name of a style made in France's Jura region. It smells like roasted hazelnuts, is the color of good French butter, and tastes similar to sherry but with less alcohol. Vin jaune is made entirely from the Savagnin grape, which is neutral and related to Chardonnay. This style of wine is remarkable on account of the way it is made, through a process of aging under a layer of naturally forming fluffy yeast called le voile ("the veil"), which forms on top of the wine thanks to the Jura's humidity. Le voile covers the wine beneath it for precisely six years and three months, at which point the winemaker bottles it. By then, the wine has become a versatile and entrancing new thing, especially if you love citrus, curry, green apple, and hazelnuts.

ORANGE

Orange wines are made from white grapes that are treated as if they were destined to become red wine. Grape skins are left in contact with the juice for a few days to a few weeks, lending tannins and a golden color before they are pressed, then aged, often in terra-cotta vessels buried underground. The orange color is a result of both oxygen exposure and pigment from the grape skins. These wines are toothsome and nutty and can be wonderfully complex. We can thank the country of Georgia, where the terra-cotta vessels are called qvevri, for inventing this style about six thousand years ago. We can also thank a handful of wineries in Italy's Friuli-Venezia Giulia region and Slovenia's bordering Goriška Brda (*gore-EESH-ka*

53

Noble Rot, or Botrytis

Botrytis, a fungus formally known as *Botrytis cinerea* and more commonly called gray rot or bunch rot, is a fuzzy gray mold that grows on grapes, strawberries, and other fruits destined for the compost bin. However, when environmental factors are just right, botrytis and grapes perform an astonishing dance, creating one of viticulture's true miracles. In this case, botrytis is called noble rot, and the grape transforms. Externally, it shrivels and turns a soft purple, the color of dusk. Inside the grape, botrytis concentrates the sugars, glycerol, and acidity and transforms the grape's flavors to variations of saffron, ginger, chamomile, and tangerine.

Conditions that produce this noble rot need to involve a near-perfect balance of humidity, sunlight, and temperature. Typically, rivers or shallow lakes are involved. Tokaji *(toak-EYE)*, a region in northern Hungary, is credited with inventing botrytized sweet wine as early as 1600. In Tokaji, the Tizsa and Bodrog Rivers produce early morning fog in the vineyards, perfect for transforming Furmint and Hárslevelű grapes.

As legend goes, a prominent landowner in the Tokaji region was away from home in preparation for a battle. His wife and vineyard manager decided to postpone harvest until he returned, by which point the grapes had shriveled with noble rot. They harvested anyway rather than toss the grapes, and in doing so they created a new category of wine.

Bordeaux's flat vineyard area between the Garonne River and its tributary the Ciron also has famously favorable conditions for botrytis. Fog envelops Sémillon grapes (as well as, to a lesser extent, Sauvignon Blanc and Muscadelle) to produce the wines of Sauternes, a golden liquid where sweet is the region's identity and pre-ferred style, dating back to the mid-seventeenth century. Château d'Yquem is the crown jewel, with a vineyard benefiting from warm air that combines with the cool water from a spring that feeds the Ciron. Thomas Jefferson was among Château d'Yquem's many high-profile admirers. Other botrytis-friendly environments include Austria's vineyards surrounding the large, shallow Lake Neusiedl, and Germany's vineyards along the Mosel, Rhine, and Nahe Rivers. The town of Rust, along Austria's shallow Lake Neusiedl, even managed to purchase its own independence from Austria's emperor with proceeds from its sweet wine sales back in 1681.

BUR-da) region—looking at you, Gravner and Radikon!—for taking it mainstream. Now great winemakers worldwide are experimenting with these techniques and making skin-contact orange wines from any white grape you can imagine.

SWEET

Sweet wines were, for centuries, the rarest and most famous in the world. Richard the Lionheart served a late-harvest wine from Cyprus—Commandaria—at his twelfth-century wedding and called it "the wine of kings and the king of wines." The wine was first referenced and made famous by the ancient Greek poet Homer and is the oldest named wine currently still in production. (You can buy it today, and it tastes like a blend of liquefied figs, prunes, and dates. It's great over vanilla ice cream.) Ancient Romans added honey to their wine to sweeten it, and only kings were allowed to drink ice wine and Trockenbeerenauslese, a botrytized wine from Germany. Botrytis, after all, is called noble rot.

Sweet wines are not all that common at dinner parties these days, which I get. Everyone is looking to decrease their sugar and alcohol consumption, and often I'm on that same bandwagon. But sweet wines can be magical, and they are more than worth having on hand for those nights when you are hosting dinner and want to extend the evening. As a bonus, you can dazzle your guests with a history lesson. Pour them a glass of the legendary botrytis-infected South African sweet wine vin de Constance while also sharing some historical insights. (Jane Austen, Charles Dickens, and Charles Baudelaire were all fans, and Napoleon is said to have requested it on his deathbed.) Or skip the insights and serve a glass anyway alongside squares of excellent dark chocolate, fresh clementines, and dates. Boom! Dessert perfection, no cooking required.

Sweet wines can be made in a variety of ways, but the most common follow.

55

LATE-HARVEST WINES: This is exactly what it sounds like: allowing grapes to become raisins on the vine long after they've ripened. The acidity has dropped, sugars have increased, and water in the grapes has evaporated, concentrating the flavor.

BOTRYTIZED WINES: Botrytis, also known as noble rot, affects individual grapes by encasing them in its fuzzy, purplish mold that changes the pulp inside by increasing sugar, acidity, and glycerol, giving the resulting wine a thicker texture. It also adds its own unique flavor, a heavenly mix of saffron, ginger, chamomile, and tangerine.

ICE WINES: These are made from grapes unaffected by botrytis that manage to make it intact all the way to the first frost long after ripening. They're harvested on the night of the first frost, then pressed immediately. Frozen water in the grapes stays behind as ice while the sweet, sugary juice is extracted, then turned into ice wine. Peller Estates in Niagara-on-the-Lake, Canada, makes terrific examples from grapes spanning Vidal to Cabernet Franc (the color of the latter is a vibrant red, like a ruby, and stuns dinner guests every time).

56

FORTIFIED

Fortified wines are like that quirky friend you can seat next to anyone at the dinner party. By design, they're shelf-stable and they've been world-famous long before refrigeration. They're also quite versatile and range from bone-dry to sweet. In the case of some sherries, they're not even boozy, an anomaly given that fortified wines are, by definition, wine spiked with booze. Think of fortified wines as your cellar's insurance policy.

Some styles of sherry make terrific aperitifs, and Madeira and ports can be wonderful cheese pairings or after-dinner options. You can keep a bottle or two of Fino sherry on hand for those days when you're spontaneously hosting aperitivo hour and forgot to pre-chill a bottle of Champagne. Port is versatile and long-lasting—and accompanied by peculiar rituals (it must be passed around a table clockwise, starting and stopping with the host). Madeira is both delicious and nearly indestructible. It is also my personal favorite fortified wine.

A SHERRY, PORT, AND MADEIRA PRIMER

At some point during fermentation, fortified wines get spiked with a clear, flavorless spirit—grape brandy, which is similar to vodka. Even the strongest yeast can't survive in that much alcohol, so the spirit kills them instantly, and the now-fortified wine is shelf-stable and can withstand temperature extremes that other wines cannot.

A brief primer is below, with some terms you're likely to see on bottles in restaurants or wine stores. These are terrific options for when you're feeling adventurous and itching to get out of your comfort zone. If this isn't your thing, no worries, just skip ahead—the majority of the book focuses on red, white, and sparkling wines, but I didn't want to omit these fortified options!

SHERRY

Sherry is a protected name for fortified wines made within the "Sherry Triangle" in Andalucía, Spain. The term *sherry* is an anglicized version of *Jerez de la Frontera*, usually known as just Jerez, the most famous of the three towns that make up the triangle. (The other two are Sanlúcar de Barrameda and El Puerto de Santa María.) Sherry comes in various styles, from pale and bone-dry to syrupy brown and molasses-sweet. The kinds that I find most interesting are dry Fino and Manzanilla sherries, which are both made from the white Palomino grape and undergo a very cool process known as biological aging, the same process responsible for vin jaune (yellow wine; see page 53).

Fino and Manzanilla are both made from the Palomino grape grown on white, calcareous albariza soil. After fermentation, the dry white wines age under a blanket of flor (a biofilm of yeast that looks like a pale angora scarf). Flor impacts the sherry beneath it in some interesting ways: It prevents oxygen exposure, halting oxidation; it eats alcohol; and it eats sugar and glycerol to boot! So, while sherry is fortified, Fino and Manzanilla are remarkably light and refreshing.

Below is a quick-reference guide to the different sherry styles.

FINO AND MANZANILLA: Both of these are made from the Palomino grape, and they're the driest and palest sherries out there. Their differences have to do with where they're made. Manzanilla is made in the coastal Sanlúcar de Barrameda, where the flor is thickest. These are the lightest, most ethereal sherries you're going to find. Fino, on the other hand, can be made everywhere in the region except Sanlúcar de Barrameda. Finos are briny, slightly nutty, and just not quite as light or dry as Manzanilla. They remind me of olives and lightly toasted almonds. (Those foods are very good pairings with Fino and Manzanilla, by the way. Don't hesitate to drink them with some bread with anchovies, while you're at it.)

AMONTILLADO: Basically, this is a sherry that started out as a Fino, but the flor didn't develop properly. The flavors are darker, toastier, and stronger than Fino and Manzanilla, with more oxidative notes (think sliced red apple that's been sitting on the counter for a couple of hours). If you have a bottle at home and are wondering what to drink it with, hard cheese, tinned fish, and wild mushrooms are all good options.

OLOROSO: This is the last stop on the oxidative train for sherries made from the Palomino grape, and the wines do not spend any time under flor (rendering them higher in alcohol). Chestnut brown in color and a glorious host of contradictions, Oloroso *smells* nutty and sweet, but it's bone-dry. Admittedly, I do not drink Oloroso often, but it could be a fun option with roasted meat when you want to switch things up from the usual full-bodied red.

MOSCATEL AND PEDRO XIMÉNEZ: Conveniently, these are the names of both the grapes and the actual wines, which are sweet and syrupy and taste like caramel, spices, and raisins. They would be terrific with something like a sticky toffee pudding, as they're essentially the liquid version of that.

EN RAMA: This term shows up on some of the dry styles and means unfiltered, or rather, filtered with a much wider filter, allowing some bits of flor to enter the bottle. Expect a bit of cloudiness and a fuller, richer flavor.

MADEIRA

Madeira is a subtropical volcanic island off the coast of Africa, west of Morocco in line with the Sahara Desert. Technically, it's part of Portugal, and it has a long, six-hundred-year history of making wines. Goods, including wines, traveled in barrels by ship throughout the world, and historically fortification helped stabilize Madeira's wines for their journeys across the sea.

Madeira's particular spin on fortification is a happy accident, resulting from barrels journeying out to sea and across the equator over a period of several months and sometimes back again. One merchant supposedly liked the returned casks more

MADEIRA AND AMERICA

Madeira, whose popularity peaked in the eighteenth century, became a particular favorite in the American colonies. The wine grew so beloved that the Founding Fathers chose it to toast the signing of the Declaration of Independence. Madeira is also said to have been, alongside walnuts, the "inevitable last course" at dinner parties of that era according to late journalist Benjamin Perley Poore's memoir—a tradition I will look to recreate!

than the original version, and a new style was born. Today that style is recreated by cooking the wine in an estufa, a steel tank with heated coils inside—far more efficient and economically viable than sending barrels on long trips across the ocean.

The island's popularity suffered several blows in the mid-nineteenth century, from powdery mildew and oidium to the root louse phylloxera. While some wineries have recovered, the island currently produces a fraction of what it used to. Most Madeira today is made from the less illustrious Tinta Negra Mole grape. Madeira's great wines, however, are still made from noble varietals that, conveniently for drinkers, are written on labels.

Below is a guide to Madeira's noble grapes, from driest to sweetest.

SERCIAL: Bright and nutty, the driest of the noble varieties. Think walnuts and dried figs with a little preserved lemon.

VERDELHO: A step up in sweetness from Sercial—more figs, candied walnuts, and prunes.

TERRANTEZ: Usually spotted in vintage bottlings, this is a rarely seen grape that can produce sweet or dry styles. It has a bright and lively acidity, even when decades old. Raisins, chocolate, prunes, walnuts—it's all in here. This is always a personal favorite!

59

BUAL/BOAL: Prunes in their juices (like the ones in breakfast buffets), candied walnuts, hints of milk chocolate.

MALMSEY: The sweetest in style. Nutty, nougaty, and chocolatey—sort of like a melted Snickers bar.

You may find some additional Madeira terms on labels throughout your travels. Here is a guide.

COLHEITA: Bottlings are made with a single grape varietal from a single year, and the Madeira is aged at least five years in cask.

VINTAGE MADEIRA: Used rarely and contains only grapes (any variety or blend) harvested in the stated vintage.

FRASQUEIRA: Vintage Madeira from one of the noble varieties (above) that ages at least twenty years in a wooden cask before it is bottled.

PORT

Port, from Portugal, owes its popularity and historic importance to seventeenth-century trade wars between France and England. In 1693, after various wine importation bans, King William III imposed a tax on French wines, driving English merchants to Portugal. In preparation for their products' overseas journeys, winemakers spiked the wines during fermentation, stabilizing them and ensuring a level of sweetness that the Brits, and soon other people, enjoyed.

Port can be red, white, or rosé (though red is the most famous), and it comes with an abundance of confusing label terminology, explained in more detail below. There are two main categories of port: Ruby, which ages in the bottle, and Tawny, which ages in a wooden cask.

RUBY PORTS

There are five subcategories in the Ruby port family.

RUBY: Bottles labeled simply "Ruby" are the least complex, least expensive ports. These are usually aged in bulk for two to three years before bottling, and they're not allowed to carry a vintage date. If we were going to use car analogies, these are like the AMC Gremlins of ports.

RUBY RESERVE: Also known as Vintage Character, these are a little more complex than a basic Ruby port. These are reliable but not very sexy—they're like the Honda Civics of the port world.

VINTAGE: Vintage port is the most expensive style of port and represents only about 2 percent of production. A port house will usually declare a vintage year only during truly great harvests. Often a given house will, on average, declare a vintage three years out of every decade. Vintage port must be authorized by the regulating government body, the IVDP (Instituto dos Vinhos do Douro e Porto). It will continue to develop in the bottle for decades. In car terms, these are terrific—the BMWs and Mercedes of ports.

SINGLE QUINTA VINTAGE: These are the products of one estate's harvest. They're made in the same manner as Vintage ports and will improve with additional bottle age. These are the Rolls-Royce of port.

LATE-BOTTLED VINTAGE: Late-Bottled Vintage port (LBV) spends between four and six years in cask before they're bottled. These wines benefit from the mellowed tones of a Tawny port while retaining the youthful fruit of a Ruby port. LBV port is always the product of a single vintage, but quality varies greatly (so we will skip the car analogy on this one).

In theory, these are ports aged in wood for so long that they lose their ruby color and their sharp, youthful tannins become tawny and gentle. In practice, a lot of producers simply work with inferior grapes from the outskirts of Porto and make low-tannin, low-quality port labeled Tawny instead. Below are some guidelines for Tawny ports.

TAWNY: Many wines labeled as "Tawny Port" don't benefit from the cask aging important to the style. They're simply paler and lower in tannin often because they begin as lower-quality grapes. The best way to identify these are through price point (inexpensive grocery store Tawny ports are not hidden values).

RESERVE TAWNY OR AGED TAWNY: Reserve Tawny port is a step up in quality. These ports age in oak casks for at least seven years before bottling, giving enough time to smooth over any residual jagged tannins.

TAWNY WITH INDICATION OF AGE: Tawny port may be labeled as "10-Year-Old," "20-Year-Old," "30-Year-Old," or "40-Year-Old." This is an approximation that corresponds to what a tasting panel at the IVDP believes a port of that age would taste like.

COLHEITA TAWNY: This is the top of the top tier, a vintage-dated port that spends a minimum of seven years in cask and can stay in cask for decades.

CHAPTER 3

FROM VINE TO BOTTLE

THE ART AND SCIENCE OF WINEMAKING

One of my favorite translation discoveries is that the term *winemaker* does not exist in France. In France, the first country to treat winemaking as an art and home to most of the world's benchmark examples, there is—at least linguistically—no ego involved. Instead, the word is *vigneron* ("vine grower") and it is a great reminder that ripe grapes, regardless of human intervention, ferment. The implication here is that a human's job is to competently grow these grapes so that nature can do its part and turn them into something even more spectacular. Humility, a tenet of every great vigneron, is literally in their job title's DNA.

Ever since the nineteenth century, however, when Louis Pasteur discovered the role of yeast and bacteria in fermentation, human intervention has become a more powerful factor. Winemaking today is a robust industry with potions, powders, gadgets, and consultants on hand to help refine grape

juice on every step of its journey to becoming wine. The English term *winemaking* sums up the philosophical difference between a modern approach to wine production (that wine must be made, like a cake, with ingredients and instructions) and the French concept of vine growing, which reveals a more cooperative and traditional viewpoint—that wine is the result of a holistic collaboration with nature.

PATIENCE

In 2012, I had the privilege of working the harvest at Domaine Georges Mugneret-Gibourg, a winery run by three generations of incredible women in the famous Burgundian village Vosne-Romanée. When Georges Mugneret passed away in 1988, he left the winery in the hands of his widow, Jacqueline, who then passed it on to their two daughters, Marie-Christine and Marie-Andrée. The third generation, cousins Lucie and Marion, have taken over care of the domaine.

Georges was an ophthalmologist by profession, and winemaking was something he loved to do in his free time. He and Jacqueline, a teacher, encouraged their daughters to pursue their own professions as well. (Marie-Christine and Marie-Andrée have degrees in chemistry and oenology—the study of wines—respectively.)

One day during the busy harvest, the winery was peaceful and so quiet I could hear the grapes fermenting in their vats. Marie-Christine and Marie-Andrée met me to share that there was nothing to do that day, and no need to overtinker. "The grapes need to ferment," they said. That was it. We needed to wait. I spent the day at Tonnellerie François Frères, where I learned about barrel-making. We returned to the winery the next day. I learned that the courage to wait and do nothing is as important a decision as any.

Intention has an outsized impact on the final wine. Some are indeed made in factories following a recipe, while others are lovingly crafted from hand-harvested grapes tended according to the moon's cycle, vinified on small farms or in garages. There are a thousand decisions in every bottle. We covered some winemaking basics in chapter 2; this chapter explores the key decisions a winemaker (or vigneron) encounters in a given year. (I'm going to use the terms *vigneron* and *winemaker* interchangeably from here on out.)

CYCLE OF THE VINE

We're going to breeze through a quick overview of the vine's growing cycle because I think it is useful to understand the factors a winemaker manages before and after harvest. While winemakers don't control a vine's growing cycle, the decisions they make can impact when and how the grapes ripen.

DORMANCY

PRUNING

BUD BURST

LEAF DROP

LEAF GROWTH

FLOWERING

HARVEST

VERAISON
LEAF THINNING

FRUIT SET

SWEET SUNSHINE

Now is a good time to brush up on photosynthesis, which has likely not occupied much of your brain space since high school. Photosynthesis is the process by which plants use sunlight, water, and carbon dioxide to create oxygen and energy in the form of sugar. For the purposes of wine, the more sunlight vines are exposed to, the more sugar ends up in those grapes. We can thank photosynthesis for that.

Grapevines are wonderfully resilient, able to thrive on the steepest slopes and rockiest soils in countries as hot as Egypt and as cold as Canada. While temperature affects the length of the growing cycle (generally warm climates have shorter cycles because grapes ripen faster; cool climates have longer ones because grapes ripen more slowly), the main growth stages everywhere in the world are the same.

These stages include budbreak in spring, flowering and fruit set in late spring and early summer, veraison in summer (when sugar in the roots travels up through the vine into the grapes and they transform from hard green pellets to red, pink, or golden globes of sweet fruit), and finally harvest in late summer or autumn. After harvest, vines are pruned, the sap in the canes returns to the roots, vines rest (or go dormant) during the winter, and the process begins again in spring.

NAVIGATING NATURE'S ANNUAL OBSTACLE COURSE

Every step of the way, vignerons and their teams manage environmental factors that have a terrifying potential to scrap their chance to harvest. Spring frosts threaten vines at budbreak in cooler regions, potentially wiping out entire crops. Hail is a danger all season long, pelting buds off the vines or shredding grape leaves and stunting growth by slowing photosynthesis, or by ruining mature grapes if the storm hits in late summer or fall. Downy mildew and powdery mildew are fungal diseases that wreak havoc anywhere with rainy springs, spreading on young leaves and buds and halting growth and ripeness.

Additionally, hungry animals (birds, wild boar, and even kangaroos) all love grapes—who can blame them? And climate change is bringing new challenges to growers everywhere. Heat waves and wildfires, once occasional hazards, are unfortunately the new norm.

HARVEST DATES AND CLIMATE CHANGE

Burgundy has kept records of the legal harvest date, or *le ban de vendanges,* since 1354 CE, and harvest dates have crept up alarmingly over the past several years. From 1354 CE until 1988, the average harvest date was September 28. In the years from 1988 to 2021, the average harvest date crept up thirteen days earlier. In 2022, the harvest started in mid-August, and harvesters were in vineyards in late August in 2023.

HARVEST

Harvest is an extremely short pressure cooker of a window when an entire year's work is on the line. Once veraison begins, grapes quickly fill with sugar, and acidity drops. If winemakers harvest too soon, they'll end up with tart, bitter grapes and unripe phenolic tannins. If they harvest too late, they'll have flabby wine that will have to be "corrected" (mixed with tartaric, malic, or citric acid) in order to bring some brightness back. In some regions, even the time of day harvest occurs is important. More and more often, harvest begins early in the morning when it's still dark, to preserve as much acidity as possible. Even regions like Burgundy and Barolo have begun harvesting in the mornings, because temperatures are too hot to pick in the afternoon.

Many winemakers today send grape samples to a laboratory for an analysis when choosing a harvest date based on reports that convey sugar levels and acidity. However, some great winemakers bring more intuition, and their own palate, to the process. Jean-Marc Roulot, a world-renowned vigneron in Meursault, walks through his vines tasting grape samples, analyzing the sweetness, acidity, and phenolic ripeness (structural maturity) in the skins, grape pulp, and seeds. I recall watching him analyze the seeds in particular—if grape pulp clung to them after he spit them into his hand, the grapes were not ready. If it did not, it was one sign the grapes were structurally mature.

67

Maisons and Domaines (or, What's in a Name?)

Wines made from the same vineyard and vintage can vary wildly in quality. Early in my career, I worked at Maison Deux Montille, a winery that was overseen by winemaker and chef Alix de Montille and her brother, Étienne. Alix and Étienne owned vineyards and also oversaw winemaking at their family estate, Domaine de Montille, but Maison Deux Montille allowed them to make more white wines, something Alix loved. A winery that makes wine from purchased grapes, juice, or sometimes even wine from the bulk market is called a négociant. When the word "Domaine" is absent on a label of Burgundy wine, that gives the winery flexibility to add purchased fruit if they so choose. The example on the left below indicates that bottle was made exclusively with grapes grown and harvested by the winery, while the example on the right indicates that may not necessarily be the case.

Alix had agreed to trade some grapes from Puligny-Montrachet Le Cailleret, a remarkable Premier Cru vineyard she and her family owned, for some Grand Cru Bâtard-Montrachet grapes owned by a big-name winery with fancy vineyards of its own. One morning, a pickup truck rolled up to the winery. A man got out and plunked several plastic cartons of bright green grapes onto the pavement, handed me a packing slip, and then sped off. When we investigated, we were stunned to find that the large winery had decided to deliver the grapes a week before the agreed-upon date, without notice or approval. Nearly half of the grapes were verjus, the second-growth grapes that no self-respecting winery harvests at all, unless they are selling them off in bulk to make cooking wine. I spent the afternoon sorting verjus from wine grapes, and we drove back in tears to the winery to deal with the broken contract. It was a wake-up call and a chance to see firsthand that famous wineries and expensive vineyards do not guarantee a good product!

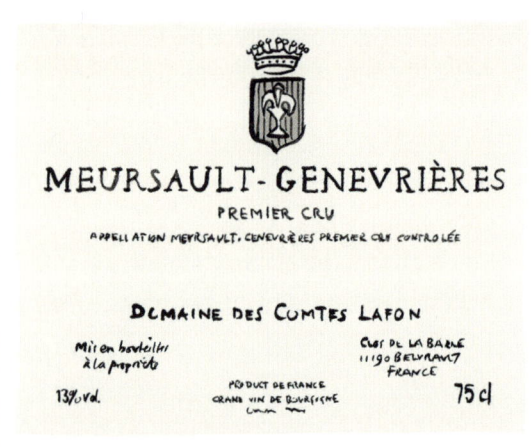

HAND-HARVESTING VERSUS THE MACHINE

Grapes can be harvested by hand or machine. Hand-harvesting is more precise, but it's also more expensive because it requires actual people—the more the better. Machine harvesting is less expensive because no humans are required, but it's less precise.

Hand-harvesting is backbreaking work. It involves people spending hours lugging around buckets of grapes, often in a crouched position, then lifting them into flat bins that can be stacked and transported to the winery, where they are sorted and processed. Most of the best estates have sorting tables—conveyor belts on which the grapes travel past focused hands that remove bugs, leaves, moldy grapes, and anything that looks unsavory. A rule of thumb: If you don't want to eat it, you won't want to drink it as wine.

Machines, on the other hand, straddle the vines, shake them, and collect everything that falls in their path. Ripe grapes are the main event, but unripe grapes, dried leaves, bugs, birds' nests, and even the occasional snake or critter make it into the basket, too. Sorting machines are getting more advanced, but they're far from perfect.

TO DESTEM OR NOT TO DESTEM

When making red wines, one decision a winery confronts is whether or not to destem. Destemming involves a machine that separates grapes from their peduncles (aka the stem or the stalk). (*Peduncle* is such a fun word to say that I'm including it here, but note that it is unlikely to pop up in conversation.) Destemming proponents argue that stems contribute an undesirable "green" or "stemmy" taste. However, some of the world's greatest Pinot Noir wines opt for whole-cluster fermentation, with nary a hint of stemminess. Advocates believe whole-cluster fermentation enhances texture, creating soft, gentle tannins from grapes that are able to begin an intracellular fermentation (where the fermentation starts to happen inside the individual grape berry), which creates tannins as soft and silky as satin pajamas. Additionally, fermenting whole clusters can add more overall complexity; intact clusters create pockets of air within the vat, and different yeast strains can develop in those individual air pockets, lending more depth and nuance to the finished

DESTEMMING VERSUS WHOLE-CLUSTER FERMENTATION

Few topics are as contentious in Pinot Noir production as that of destemming. The late Henri Jayer, a proponent of destemming, proclaimed, "No great wine has ever been made using whole cluster," at a dinner that included Aubert de Villaine of Domaine de la Romanée-Conti, arguably the most famous winery in the world today, and a whole-cluster fermentation poster-child estate.

wine, sort of like a soup flavored with several herbs (as opposed to just one). Even the gentlest destemming machines invariably tear the grape berries, which then begin to release juices and get squished beneath the weight of grapes that fall on top of them in the vat. The result is a sort of grape slushie, devoid of air pockets for different yeast strains to grow, and no full berries to foster intracellular fermentation or the silky tannins it promotes.

PRESSING AND FERMENTATION

Once the grapes are harvested, the winemakers are in a sprint to make smart decisions as quickly as possible, all day long, for several weeks—while also employing and feeding a team that is exponentially larger than their usual team, thanks to the influx of temporary harvesters.

BLADDER PRESS

WOODEN PRESS

COQUARD PRESS

Grapes destined for white wine are pressed as quickly as possible after they've been picked, ideally within the hour but not longer than a couple of days. Then the freshly pressed juice, called must, is moved (usually by a pump, but also sometimes by gravity in wineries that want to be especially gentle) to a tank, where the grape must ferments into wine. Winemakers have to act quickly as any corrections or additions are much more challenging once fermentation is under way.

Grapes destined for red wine are transferred to a fermentation vessel, where the winemaker monitors sugar, acidity, temperature, and other factors to determine if the yeasts are happy and healthy. In addition, winemakers must consider color and tannins and extract a desired amount of both. Tannins have a small impact on flavor (they add a slight bitterness) and a large impact on texture. They even have a small impact on a wine's longevity. Managing tannins is no small task!

Below are some of the main decisions a winemaker encounters while must ferments, during that crucial moment after harvest and before élevage (aka raising the wine until it's ready for bottling).

TYPES OF PRESSES

I had no idea the type of press could make such a difference until visiting so many wineries with strong preferences for each one. Three types of presses are most commonly used in reputable wineries: the bladder press (or pneumatic press), the Coquard press (most typical in Champagne), and the traditional wooden basket press. Below is an overview of each.

1. **BLADDER PRESS:** Also called pneumatic presses, bladder presses are perceived as a more delicate method for extracting juice than traditional wooden presses. They are also far less labor-intensive. An invention of the 1950s, the press often resembles a giant stainless steel vitamin capsule. Grapes are gently dumped into the top, where they lie next to a deflated bladder. The winemaker programs the machine and the bladder fills with air, pressing the juice through a strainer at the bottom and into a tub that collects it below. Some of my favorite wines in the world are made using a bladder press—as are some far more lackluster options. Precision and thoughtfulness in programming the press seem to be determining factors.

2. **COQUARD PRESS:** The Coquard press is especially beloved in Champagne. I spoke with Raphaël Bérêche, a very talented Champagne producer who loves his Coquard press, about why it is noteworthy. He broke it down for me as follows:

> Unlike vertical (wooden) presses, the Coquard is a horizontal press with a diagonal edge that flows down, like a slide, into the tray collecting juice, below. This diagonal angle at which the juice is pressed is critical for a couple of key reasons. First, the suspended proteins in the grape juice flow

directly down into the tray, carrying flavor and complexity with them. In other presses, such as the vertical wooden press or the bladder press, the juice must first pass through a wall of pressed grape stems and skins, where many of the proteins (and flavor) get trapped instead of flowing down to the tank. Additionally, the diagonal angle creates a natural pressure to guide pressed grape skins up, in a circular motion. This allows the heavier, juice-filled grapes to fall to the bottom for pressing, where the juice and proteins are extracted and fall into the tank.

3.   **WOODEN PRESS:** Old-school, traditional wooden basket presses are coming back into fashion in several wineries. Aside from the fact that they look cool and vintage, they serve an important purpose in extracting a lot of juice and phenolics (chemical compounds responsible for both flavor and tannins, among other things). One downside: They're manually operated and require a lot of effort! Someone churns a crank, forcing a wooden top to press down on the grapes, extracting as much juice as possible out of the sides of what is essentially a large wooden basket. In recent decades, with premature oxidation (also known as premox) occurring in white wines throughout the world, particularly in white Burgundy, some people are returning to wooden presses because they believe bladder presses are too gentle and don't extract enough tannins (which add texture to the wine but also act as a preservative).

CAP MANAGEMENT!

In red or orange wine, winemakers have to manage tannin and color extraction during fermentation. This involves deciding how to handle the "cap," or the mass of grape skins and seeds that floats to the top of the fermentation tank.

Three techniques are most common:

1.   **PIGEAGE** involves submerging the cap beneath the juice regularly, usually with a tool that looks like a giant plunger. This technique gives a lot of control to the winemaker, who can decide how vigorously and frequently to pigeage and therefore how aggressively or gently to extract tannins.

2.   **REMONTAGE** requires pumping grape juice from the bottom of the tank over the top of the cap to keep it wet, circulate oxygen, and gently extract tannins.

3.   **DÉLESTAGE** entails racking (removing) all of the fermenting grape juice off the cap and transferring it into a separate tank, then returning the juice to the original tank with a hose, thereby adding lots of oxygen along with plenty of force. According to advocates, this technique softens astringent tannins by essentially pummeling them.

MANAGING FERMENTATION

Fermentation is a critical time for wine because the must is at its most vulnerable. Stuck fermentation—when yeasts stop eating sugar before fermentation is complete—is the ultimate fear, because it makes the wine susceptible to spoiling. In stuck fermentation, bacteria in the winery can cause volatile acidity (VA) or other volatile compounds—usually acetic acid (vinegar) or ethyl acetate (nail polish remover)—that are devastating. (See page 38 for more information on VA.) Temperature control in a winery is one way to help mitigate this ultimate fear.

MALOLACTIC FERMENTATION

In wine, there are two important fermentations: alcoholic and malolactic. First, during the alcoholic fermentation, sugar is converted to alcohol with help from yeasts. After the alcoholic fermentation, wines can undergo malolactic fermentation, sometimes called the secondary fermentation (even though it's not technically a fermentation) or malolactic conversion. During this phase, bacteria convert the malic acid (think tart Granny Smith apple) to lactic acid (think tangy yogurt). This secondary fermentation stabilizes the wine, decreases overall acidity, and produces diacetyl, the chemical compound that tastes like melted butter and gives wine a richer, rounder mouthfeel. This is common in most red wines and some whites, notably Chardonnay. Sometimes malolactic fermentation happens naturally; at other times the winemaker controls it, inoculating the wine with lactic bacteria. Malolactic fermentation can also be "blocked" in white wines by winemakers who want to keep that crisp, malic acidity.

SULFITES AND OTHER ADDITIVES

Back in 1986, the US Food and Drug Administration (FDA) identified sulfites as an allergen based on several asthma cases. Wineries were getting heavy-handed with sulfur, using it as a crutch to fight bacterial infections and other issues in the winery (without just keeping the winery cleaner!). Starting in the late 1980s, sulfites had to be disclosed on wine labels if detected at 10 mg/l or higher, or if any

ALCOHOL IN THE WILD

While the following info may seem random, I'm adding it as a reminder that grapes—and fruit in general—ferment on their own. It turns out that a lot of insects and animals drink alcohol in the wild, thanks to naturally fermented fruits and grains. According to *National Geographic*, male butterflies consume alcohol (beer is preferred) to "boost their spermatophores." Yes, just what it sounds like. Unmated fruit flies seek out alcohol to increase a brain chemical called neuropeptide F, which gives them a feeling of pleasure and satisfaction that mated fruit flies find elsewhere (who knew?). Moose, squirrels, and white-tailed deer get tipsy, too. Bats consume especially large amounts of alcohol but apparently have tolerances so high that their flying is not impaired!

Manipulations and Corrections

These not-particularly-flattering terms refer to the range of interventions winemakers can use to guide and refine a wine from harvest through bottling. Below is a list of some of the most impactful manipulations in the order they occur. Wine can be tweaked with everything from tartaric acid to sugar to nitrogen, all in an attempt to get the best balance of acid, sugar, and yeast for each stage of the wine's production.

Acid Adjustment

A nicer way to say "acidification" or, in rare cases in cool climates, "deacidification." Acidification is the process of adding acid to must or to fermented wine. Typically, this happens before fermentation has finished, as adding acidity to must makes a more stable wine and is less likely to add funky aromas. Acidity is a miracle worker in that it helps protect against bacterial infections and certain yeasts like *Brettanomyces*, and it even helps preserve color. Tartaric acid is the most commonly used for a few reasons—it's the naturally occurring acid in ripe grapes, and unlike malic and citric acids (the other two options), it won't get attacked by lactic acid bacteria.

Barrel Fermentation and Oak Aging

The size and age of a barrel impact the taste and texture of a wine in several ways. New oak gives flavor, but oak's flavor-infusing powers decrease with each vintage. Smaller new barrels lend more flavor than larger ones because a higher amount of the oak's surface area is in contact with the wine. Oak is porous, which means oxygen flows into and out of the barrel. However, the barrel age impacts this flow. New barrels allow more oxygen flow, while older barrels allow less—their pores are clogged with wine and tartrate crystals (naturally occurring solid bits of tartaric acid) that precipitate out and impact airflow over time. For more on oak, see page 79.

Chaptalization

This practice, named after the late Jean-Antoine Chaptal, a French chemist and Napoleon's minister of agriculture, involves adding sugar to unfermented grape must to increase the alcohol (and boost flavor and texture) in the final wine. Chaptalization is historically common in cooler climates and gained traction in the early nineteenth century, apparently when Chaptal was tasked with finding a use for an excess of sugar beets. Some of the world's greatest wines use this technique, though it's becoming less common with increasing global temperatures. While Chaptal often receives credit for this technique, the Romans were adding honey to fermenting grape must centuries earlier!

Fining (or Clarification)	The process of removing solids from the wine. This can be done a few different ways and is sort of like making consommé, if you've ever done that. Essentially you mix the wine with a binding agent—like egg whites or isinglass (gelatin from a fish swim bladder), which is how some wines become non-vegan, or you can use a vegan alternative like bentonite (a type of clay). Stir this vigorously and wait for the binding agent to attach to the solid particles; the solids will bind together and be easier to remove, leaving you with clear wine.
Filtration	Filters come in all shapes and sizes, and removing particles in wine ranges from just letting the wine settle and racking the clear wine off the lees to more invasive forms, like sediment filtration (filtering the wine through layers of soil particles, like a home water filter) or surface filtration (filtering the wine through a fine-mesh strainer, like making coffee). The purpose, in addition to just making a clearer wine, is to ensure things like yeast, bacteria, grape cellulose, proteins, pectin, and even dirt don't make it into the final bottle. Reverse osmosis is a type of filtration, too (more on that below).
Lees Stirring	Also called bâtonnage, this procedure involves taking a wooden baton and whisking it around vigorously in the barrel, as though you were whipping an enormous batch of butter. Why go to this trouble? It incorporates the lees into the wine, boosting richness and flavor.
Racking	This is the process of siphoning wine off the lees. A winemaker's decision of when and how often to rack has a large influence on the wine's final taste. Some winemakers leave the wine in contact with the lees almost until bottling, as high-quality lees can add life, texture, and stability to a wine. By contrast, "dirty" lees or lees that are not high-quality lead to stinky wine if the wine isn't racked.
Reverse Osmosis	This is a type of crossflow filtration where the liquid flows parallel to the membrane filter. It's usually used for alcohol reduction and must concentration (which comes in handy for making fuller-flavored must in a rainy, dilute year). Essentially, reverse osmosis separates the wine into two parts—water and alcohol, which winemakers can tinker with depending on their objectives. In recent California vintages, reverse osmosis has been used to reduce smoke taint. Often, wineries with hygiene issues use it in an attempt to remove *Brettanomyces* from their wines.
Saignée	This technique for color concentration is most commonly used in years when rainstorms dilute grapes before harvest. Its highest purpose is to concentrate color and flavor in the must of a soon-to-be red wine. A by-product of this process is high-quality rosé that the winery can sell or consume. (See "Rosé," page 47.)

75

After I had worked a combined five harvests in Italy and France, Robert gifted me an online winemaking class at the University of California at Davis for Christmas. This elated me as few gifts before ever had, until the coursework began, and I noticed a stark disconnect between information being taught and information I had previously learned through experience. No reading between the lines necessary—only a completely different version of reality presented as fact.

For starters, the course proclaimed that "natural yeasts" capable of producing sound wines no longer exist and that yeast had to be purchased to make a "stable" wine. The video in my course even took students on a virtual tour through a winemaking shop to become familiar with various yeasts from which one might choose. The rest of the course continued—promoting potions and products, refuting information I'd personally seen or experienced, and teaching winemaking as a manufactured product instead of an artistic or agricultural one. It was as though the yeast company had paid for that section of the course. The rest of the course continued in this manner. Hopefully someone has revised it, but I am still triggered by all that propaganda and grateful to know from previous experience that the emphasis on products and technology is only one way of making wine, and it's not universal!

additional sulfites were added beyond those that occur naturally during fermentation. All that is terrific—and very reasonable.

That said, I'd argue that sulfites have been unfairly maligned. Because they are mandatorily disclosed on the label, unlike many more frightening ingredients often added but not disclosed, they're getting blamed for things like headaches, hangovers, and allergies that are actually not their fault! Don't get me wrong—I'm not a sulfur apologist. But sulfites, or sulfur dioxide (SO_2), occur naturally in fermentation. Moreover, they have been used as a preservative in winemaking since the days of the Roman Empire when winemakers burned sulfur candles in wine vats to prevent wine spoiling into vinegar. Sulfites also occur naturally in volcanos and egg yolks, and in case you have a sulfur allergy, note that there is more SO_2 in a few dried golden apricots or a tablespoon of ketchup than in most bottles of good wine.

Sulfur's mandatory disclosure on labels seems to have evolved into a way for less-quality-minded wineries to get away with adding other things—like the neurotoxin Velcorin, which is so toxic it can cause death if consumed within forty-eight hours of being administered. Other additives include Super Smoother (a mix of glycerine and liquid oak), Mega Purple (a mix of 70 percent cane sugar and Rubired grape juice concentrate), powdered tannins, powdered acid, defoaming agents, and more. The list is endless.

ÉLEVAGE

The term *élevage*, used in wine production, is also the French word that describes nurturing children into adults—it carries the same meaning for both wine and humans. It is an all-encompassing word that broadly means "raising," and it refers to the time when wine is evolving, in preparation for bottling. It also refers to the love, attention, and occasional intervention required to take fermented grape juice from its infancy into a spectacular bottle, ready to drink. Élevage can occur in stainless steel tanks, oak barrels (of various size), and cement or clay vessels, and can range in duration from several months to several years. High-tannin red wines like Nebbiolo, Sangiovese, or Cabernet Sauvignon benefit from—and are in many cases required by law to include—years of élevage, where the wines interact with tiny amounts of oxygen through the oak and the tannins soften. Below is an overview of some decisions a winery makes about fermentation vessels during this important time.

WHY FERMENTATION VESSELS MATTER

Piero Incisa della Rocchetta, proprietor of Bodega Chacra in Patagonia (and also of Sassicaia family fame—the world's first Super Tuscan wine and now its own appellation in Italy), once speculated to me that a perceptive wine drinker can actually feel the energy of a wine based on the vessel in which fermentation and élevage occurred. It struck me as far-fetched in the moment, but as I've paid more attention to this, I actually agree. A zippy New Zealand Sauvignon Blanc can taste like it is ricocheting off of my tongue and bouncing around my mouth in a way that sound reverberates off of a stainless steel tank. Many of the wines I find most grounded and supple tend to have been fermented or at least partially aged in cement. Particularly round wines were aged in an egg-shaped cement vessel, which has no corners and encourages perpetual movement of the wine throughout élevage. Wines fermented in oak casks have the benefit of spending time in a porous environment in which oxygen flows and a wine's texture deepens, and depending on the age of the wood, these can impart tannins or flavor to the wine.

The following is a list of vessels used during élevage.

OAK BARRELS: Oak barrels are used for both fermentation and élevage. Oak is porous, allowing tiny amounts of oxygen to flow into and out of the barrel. The newer the oak, the more porous the barrel. (Oak expands when it gets wet.) Small, new barrels will impart more "new oak" flavor than large new barrels because more of the wine is in contact with the oak.

OAK BARREL   **CONCRETE "EGG"**   **STAINLESS STEEL TANK**   **CLAY AMPHORA**

STAINLESS STEEL: This is a sterile, nonporous neutral material that adds neither flavor nor textural richness to the wine. One benefit: Winemakers don't have to worry about pesky rogue yeasts like *Brettanomyces* getting trapped inside, so stainless steel is a great choice for hygienic, neutral winemaking. Typically, the world's most complex wines do not spend much of their élevage in stainless steel.

CEMENT OR CONCRETE: These vats are considered semiporous, meaning they allow some flow of oxygen in and out, though less than oak. Some vessels are shaped like eggs, and these offer the benefits of concrete tanks with the additional benefits of suspending the lees, because the tank has no corners and the wine inside is constantly in motion, thanks to fermentation.

CLAY AMPHORAE: Clay (terra-cotta) amphorae, also known as qvevri or kvevri, talha in Portugal, and pithos in Greece and Sicily, are ancient oval-shaped vessels that originated in modern-day Georgia (the country bordering Russia, not the Peach State). Like cement, vinification in clay provides a happy midpoint between steel and oak. Clay is neutral and semi-porous. The ovoid shape of the vessel is thought to be ideal for fermentation because it allows the wine to circulate without knocking into any corners or hard edges and keeps it in contact with suspended lees, adding richness and texture.

RESIN, FIBERGLASS, OR EVEN PLASTIC: No romance here, and it's unlikely a winemaker will tell you if their wine was fermented in a plastic or resin vessel. I add this because occasionally, say, in a higher-yield vintage or a winery short on cash flow, a winemaker will run out of steel or concrete tanks or just not be able to afford them yet, and they'll want to ferment the grapes anyway. Sometimes a winemaker might use an extra fiberglass tub that's on hand, and everything will be fine.

All About Oak

Oak takes a while to wrap your head around when you're learning to taste. It brings flavor and tannins to wines aged in it, and these qualities differ based on everything from the species of oak (French and American are different families), how the staves are split, how long the staves are aged, and how much the barrel is toasted. At its best, oak is meant to enhance wine, not overpower it. Aubert de Villaine, co-proprietor of the world-renowned Domaine de la Romanée-Conti in Burgundy's Vosne-Romanée, considers the oak barrels he uses "a factor of purity for the wine." Oak "brings tannins which, if the vinification and élevage have been correct . . . [are] not detectable by tasting but [add] to the finesse," according to an article written by Peter Weltman and posted on Guildsomm.com.

"Oak" as a flavor descriptor in wine had a rough entrance into the twenty-first century. Supermarket "oaky Chardonnay," which my mom drank throughout my childhood and which really dinged oak's reputation, was made with "oak essence"—sort of like cheap vanilla extract, but for wine. A French oak barrel, the gold standard in winemaking, is made from a particular species of French oak that is grown, harvested, seasoned (aged), and then toasted to perfection by a set of highly skilled artisans. But French oak barrels are pricey—approximately $1,250 apiece.

New French oak adds a sweet, vanilla-and-spice flavor with notes of cinnamon, nutmeg, and clove. American oak is different; wines aged in it smell of maple syrup, coconut, dill pickles, and sometimes banana cream pie. If you are a bourbon drinker, you know the scent well—by law, bourbon must age in 100 percent new, charred American oak.

There are a few reasons French and American oak contribute such different flavors to wine. First, they're different species of oak. French oak is the *Quercus petraea* species. These trees make tighter-grained oak, which in turn result in a slower but better incorporated transfer of flavors. American oak is the *Quercus alba* species, which is less porous. One benefit of American oak is that it can be cut in more ways than French oak. French barrel maker Nicholas Keeler explains that oak has to be cut "along sap channels so that it doesn't leak," and the way those channels form in French oak is different than in American. French oak must be split; American oak may be quarter-sawn. Only about 25 percent of a French oak tree can be used for a barrel, while 50 percent of an American oak tree can be used.

Slavonian oak is the preferred vessel for northern and central Italy, on account of its tight grain and ability to produce large vessels, thereby decreasing the surface-area ratio of wine to oak and making these vessels both long-lasting and essentially neutral in flavor or structure, especially after a few years. Hungarian oak is the *Quercus frainetto* species and contributes similar flavors to French oak.

French Oak Forests

Forests are among the many things France is good at managing. In fact, 32 percent of France's forests are deliberately managed for oak! Historically, oak barrels were vessels for transporting nearly everything—and France needed a steady supply of oak for barrels as well as for the navy, whose ships were made from wood. Below is an overview of different types of French oak. Quality in French oak is generally determined by the age of the oak tree (the older, the better) before it is harvested for its oak staves.

Allier

One of six main French oak forests that provide oak for making barrels, the Allier is located in the center of France, west of Burgundy. Allier produces very tight-grained wood of the *Quercus petraea* species. The trees grow tall and straight with tight spacing, and their wood produces barrels that create soft, balanced tannins and notes of cinnamon, clove, and toffee.

Bertranges

This forest is located in central France, just southeast of Sancerre across the Loire River. As France's second-largest oak forest, it's as well known for gorgeous scenery and wooded hikes as it is for oak. Dating back to the twelfth century, the forest was gifted to a priory of Benedictine monks who lovingly tended it until it was seized during the French Revolution. Some of France's oldest and most prized oaks are grown here. Bertranges oak is known for its delicate tannin structure.

Limousin

These forests are in a hilly region north of Bordeaux that tends to produce *Quercus robur* with a wider, looser grain, giving it a stronger flavor and more tannins. This oak is usually used for Cognac aging and some Chardonnay.

Nièvre

This forest is located northeast of Allier and Tronçais, directly west of Burgundy. Gently rolling hills support tall, straight trees. Soils are primarily silica and clay (soils in oak forests are important for terroir, too). Wood from these trees tends to be very tight-grained and the resulting wines quite nuanced.

Tronçais

Within the northern part of the Allier forest is a special section of oak trees called Tronçais that are particularly tight and fine-grained, known for producing luxury barrels. (Tronçais is a subsection of Allier, but it counts as its own forest.) The trees have been cultivated since the late seventeenth century to ensure a continual supply for the French Royal Navy. This kind of oak lends a luxurious texture and mouthfeel to wine, with a slight perceptible sweetness.

Vosges

This forest is located in eastern France surrounding Alsace, bordering Germany, and due north of Switzerland. Oak from this forest tends to impart more robust tannins and deeper, darker, more opulent notes to wine. It's a great match for winemakers with a flashier style.

Oak Forests Outside of France

The following are a few other types of oak commonly used for wine barrels.

Slavonian Oak—Eastern Europe — In the nineteenth century, the very tight-grained Slavonian oak was among the most sought-after wood for large barrels and oval vats, especially by producers in northern Italy. But Slavonian oak has taken a backseat in recent years to French oak, which is now the standard-bearer for fine wines and most wines around the world—a result of terrific marketing on France's part! Italian producers have always favored Slavonian oak for their Sangiovese and extremely tannic Nebbiolo-based wines; in larger Slavonian oak casks there is less contact between the majority of the aging wine and the oak—hence, less tannin exchange, with almost no tannin exchange in large vats that have been reused year after year. Wines aged in these types of barrels tend to show fruitier notes when barrels are new.

Hungarian Oak — This oak comes from hillside forests in Slovakia, Romania, and Hungary. These barrels offer a great deal of structure and contribute tannins more quickly than French oak. Hungarian oak costs about half the price of French oak. "Hungarian oak" just doesn't have the same ring to it, I suppose!

American Oak—Minnesota and Wisconsin — Most of the American white oak used for making wine barrels comes from Minnesota and Wisconsin. American oak has long been popular with winemakers in Spain's La Rioja region and with Australian producers of Shiraz for the full, rich flavors it imparts—namely coconut—but also because it's about half as expensive as French oak, costing around five hundred or six hundred dollars per barrel.

American Oak—Oregon — Similar in flavor to French oak, this is becoming an increasingly popular choice for conscientious Willamette Valley growers mindful of the carbon footprint required to ship heavy barrels overseas. (Also, local oak is much cheaper!)

LOAM

SAND

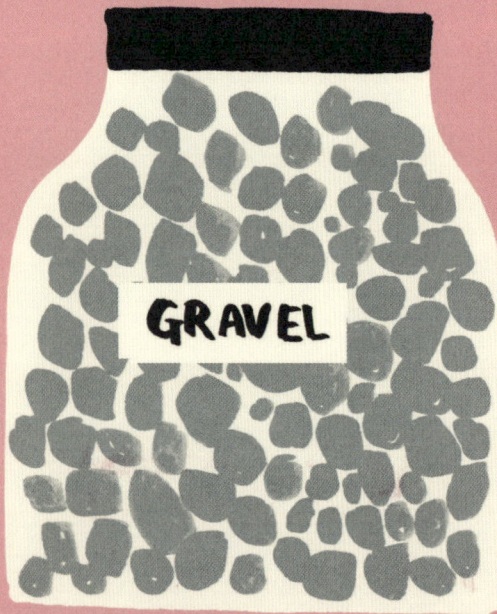

GRAVEL

LIMESTONE

GRANITE

CLAY

VOLCANIC

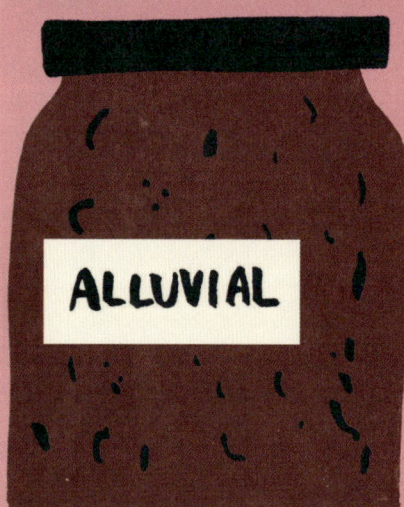

ALLUVIAL

SLATE

CHAPTER 4

LAY OF THE LAND

Land is such an immense topic in relation to wine that it almost feels unfair to smoosh it all into one chapter. Think of this as an overview of all the land-related factors at play that help shape a wine's identity. In this chapter, we'll touch on everything from getting the most out of maps to the ways in which climate factors, rocks, soil types, and farming practices all influence a bottle of wine. My hope is to add some color and context to the multitude of ways land impacts the wines you love most so that you can replicate that sense of joy whenever you feel like it.

One word that most neatly describes this impact is *terroir,* which, in the world of wine, applies to details as minute as microbes in the soil and as massive as mountain ranges influencing a region at large. Terroir is ascribed to aspects as tangible as rocks in the vineyard and as elusive as the vigneron's intention when cultivating the vines. We'll spend time on this concept in a few pages as we begin to zoom in. But first, maps help you get a conceptual understanding of how geography and climate shape your glass of wine.

MAPS ARE EMPOWERING!

As a child, the virtues of maps were lost on me, but the longer I continue my journey with wine, the more I understand their value. If you agree that a picture is worth a thousand words, maps are essential. They are your visual express train to understanding what to expect in a bottle of wine. They allow you to zoom out and get important, big-picture information that's lost when you focus on details like percent breakdowns of Petit Verdot in your Bordeaux blend. I also love maps because they bring a sense of logic and predictability to wine, a field that can often over-index in romantic descriptors and poetic language (especially from people trying to sell you bottles that don't have much else going for them). In short, maps are empowering.

A map of the Loire Valley's wine regions, for example, will show you that Sancerre (*sohn-SEHR*) and Menetou-Salon (*men-uh-too Sa-LON*) are next-door neighbors. Sancerre has the brand-name recognition, but if you're looking for something similar in style (and less expensive), put Menetou-Salon on your radar. Quincy (*kon-SEE*) across the river makes exclusively white wines from Sauvignon Blanc and is likely quite similar to your chilled glass of (white) Sancerre.

In the coming pages we'll cover a few categories of maps I find particularly helpful, but these are just a starting place. If you're getting serious about wine or find yourself smitten with a bottle or a region, maps are an incredible way to anchor that information in your brain. Whether you're a look-it-up-in-the-moment-online kind of person or have a notebook of hand-drawn maps at home, there's no wrong way to get familiar with the lay of the land.

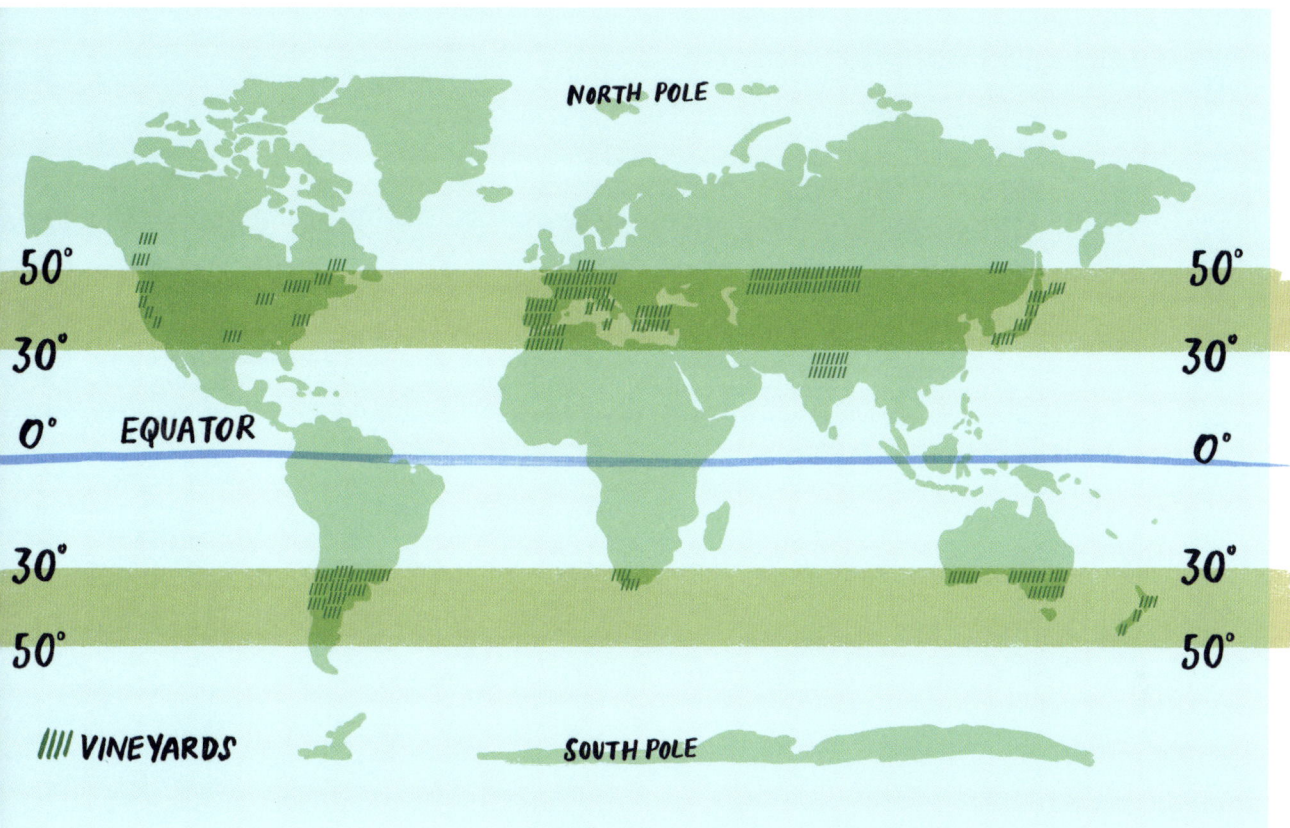

NORTH POLE

50°

30°

0° EQUATOR

30°

50°

//// VINEYARDS   SOUTH POLE

50°

30°

0°

30°

50°

GEOGRAPHICAL MAPS

Generally speaking, wine grapes thrive between 30 and 50 degrees latitude, both north and south of the equator. Look at a map of the world and you can see why vines grown closer to the North and South Poles will be lower in alcohol and higher in acidity—because these are cold places, and it's harder for grapes to ripen. The warmer the region, the riper the grapes and the higher the sugar content—and thus the more alcohol in the wine.

Maps also show major geographic factors, like mountains or bodies of water, which all play a role in the broader climate. Water in particular—the primary component of all life—impacts both macro- and microclimates when it comes to grapevines. Water both warms and cools more slowly than other substances, so oceans, lakes, rivers, and even streams all have profound influence on a region. The Mosel River, in Germany, located at 50.3 degrees latitude, falls north of the limit where vines generally grow. However, grapes are able to ripen here because the Mosel snakes through the hills, moderating temperatures. As an added bonus, the river acts like a giant mirror, reflecting light up onto the vine—a photosynthesis supercharger!

Water's absence in landlocked regions is notable, too. In continental climates, which are inland and not near large rivers or bodies of water, the absence of oceans, lakes, rivers, and streams is what creates those prized diurnal shifts (or warm days with cool nights), ensuring ripeness as well as the acidity so beloved in wines.

TOPOGRAPHICAL MAPS

This is veering into wine-geek territory, but a basic understanding of regional topography can go a long way in predicting what to expect in a bottle. Elevation has a major impact on wines, usually guaranteeing acidity from cool nights and richness from sunny days. In regions like Mount Etna in Sicily, Mendoza in Argentina, or Friuli-Venezia Giulia in Italy, elevation plays a determining role in the wine's identity.

Mount Etna towers above the rest of Sicily, often with a snowcapped peak just miles from sunbathing beachgoers. Wines from its steep, volcanic slopes grow at a distinctly higher elevation than the rest of the island. Northern Chile, at 17 degrees latitude, is far closer to the equator than other grape-growing regions, yet wines from there are all about tart, barely ripe fruit, because the grapes that produce them are grown at some of the highest elevations in the world for vine growing, at more than 6,500 feet above sea level. Because of this area's high-elevation cool nights, and because grapes are able to easily ripen thanks to intense UV exposure during the day, the resulting wines are full of zippy acidity and balance.

ZOOMING IN: TERROIR

Properly understood, it means the whole ecology of a vineyard: every aspect of its surroundings from bedrock to late frosts and autumn mists, not excluding the way a vineyard is tended, nor even the soul of the vigneron.
—James E. Wilson, *Terroir*

Now that you have a solid conceptual understanding of maps, it's time to dive into the more practical application—the land itself. Terroir is the best way I know of to begin to explain this. *Terroir* is a term that encompasses the physical—and ephemeral—details contributing to a wine's taste and structure. Terroir includes the way grapevines are pruned and trained, and whether—and how—the winemaker intervenes in vineyard management, with decisions like de-budding or green-harvesting the vines (removing extra grape bunches to concentrate flavor in those that remain). Most of the time, however, terroir is associated with a sense of place. The root word for *terroir* is the French *terre* ("earth"), and the vineyard and all its features are part of it. Terroir includes topsoil and bedrock, a slope's aspect, the way in which sunlight hits the vineyard, and even the insects that live in the vines. Terroir is also associated directly with soil.

It's worth taking a beat to reflect on our collective understanding of soil's importance. While myths that Cistercian monks tasted the dirt to determine vineyard quality have largely been dispelled, those monks still cared enough about environmental nuances to pay attention to them, demarcate them, painstakingly catalog them, and even officially name each parcel of land (known as *climats* in Burgundy—so relevant they now collectively comprise a UNESCO World Heritage Site), often after rocks. If you've ever wondered why so many villages have a vineyard named Perrières or Caillerets, well, you can thank the monks' appreciation of them. *Pierre* means "stone," *caillerets* means "pebbles," and there are dozens of other variations on the word *rock* in France, just as there are dozens of words for "cake" or "cookie" at any given Parisian patisserie.

I lean a lot on Burgundy examples in this section, both because it's where my own wine education began—and because Burgundy's appreciation for land and terroir has gone on to inform and influence the global understanding of land in relation to wine. Vineyard classification systems in countries like Germany, Spain, Italy, and beyond take inspiration from Burgundy's example—even regions that don't prioritize Chardonnay or Pinot Noir.

Now is a great time to note that people—occasionally, even winemakers I like and respect—will pooh-pooh the idea that soil and rocks affect a wine's taste, but their point of view does not resonate with what I've tasted time and again. And winemakers who make this argument tend to work with grapes grown on soil that no one is raving about.

87

LOCATION, LOCATION, LOCATION

Back in the Middle Ages, soil type and slope aspects were the critical determining factors in wine. My favorite myth on this subject is a legend about the Holy Roman Emperor Charlemagne, who supposedly had a very long white beard. He loved wine and preferred to drink red Burgundy, but as his hair whitened with age, his wife insisted he drink white wine so as not to go around embarrassing himself with a long, red-stained beard. When deciding where to plant Chardonnay grapes for his wine, he looked out to the hill of Corton and noticed a large patch where sun had melted the snow, then proclaimed that the grapes should grow there. That vineyard remains the famous Grand Cru "Corton-Charlemagne." This concept remains true today in any great wine region—slope aspect, drainage, and position all are crucial to why some vineyards are more prized than others.

SOIL, WHICH COMES FROM ROCKS

As explained by my favorite childhood scientist, Bill Nye the Science Guy, "When rocks break down into smaller and smaller pieces, they become soil. The pieces of rock mixed together with decaying plants and animals give new plants a place to grow and give earthworms, insects, and the tiny decomposer organisms a place to live." While it's a topic we're all collectively still wrapping our heads around, rocks in a vineyard directly impact a wine's texture and flavor, thanks to the ways in which they erode and interact with microbes in the surrounding soil and water. That interaction miraculously manifests as taste in your glass.

The term *rock* means a combination of minerals that come together in solid form through prolonged time, pressure, heat, or water. The kind of rock that you get in a particular place depends on circumstances that occurred there. There are three general types of rocks: igneous, metamorphic, and sedimentary.

Igneous rocks are formed through heat. Examples include basalt, which is volcanic rock formed from previously molten lava, and granite, which is molten rock that never erupted but stayed underground and burbled up through different layers of rock, mixing with them and forming a new one entirely.

Metamorphic rocks are those in which crystals have readjusted themselves. This happens at a tectonic level when plates shift (or collide) or rocks cave underneath pressure. Their atomic structure gets rearranged with all of the strain, producing a new kind of rock. Examples include schist, gneiss, and slate.

Sedimentary rocks are formed by sediment basically piling up and turning into rock over a really long period of time. Limestone and sandstone are great examples. Limestone is formed by layers of calcareous marine creatures—oysters, mollusks, sea urchins—in mud, fossilized over millions of years. Sandstone is formed when sand particles are compressed and cemented together over a similar timeline.

THE ROCK CYCLE

WEATHERING + EROSION

TRANSPORTATION + DEPOSITION

LAVA

IGNEOUS ROCK

SEDIMENTATION

UPLIFT + EXPOSURE

COMPACTION

SEDIMENTARY ROCK

MELTS

HEAT + PRESSURE

METAMORPHIC ROCK

MAGMA

Three Types of Bedrock

Igneous Rock

	BASALT / VOLCANIC ROCK	GRANITE
Description	This is terrific for growing grapes because it is porous, which means it has good drainage and grapevines can grow roots deep into the soil, where they get nutrients and water.	Formed under the earth's crust when quartz mixes with cooling magma without exposure to oxygen, this rock has a high pH that gives complex aromas and acidity to wines.
Example Regions	Santorini, Greece; Sicily (Mount Etna) and Calabria, Italy; parts of Napa Valley, California; Willamette Valley, Oregon	Cornas, (Northern Rhône Valley) France; Rías Baixas, (Galicia) Spain
Grapes That Thrive	Assyrtiko, Nerello Mascalese, Falanghina	Syrah, Gamay, Cabernet Franc, Albariño

Metamorphic Rock

	SCHIST	SLATE
Description	Schist is a hard, dense rock layered with minerals. It's usually flaky and retains heat well, and it tends to produce fleshy wines with a luxurious taste and a broad mid-palate.	This plate-like rock forms when clay, shale, or silt is pressured from within the earth. Some of its admirable qualities include its ability to warm up quickly and retain heat. It also has excellent drainage properties. Slate is found mostly in Germany, as well as in some parts of Australia and in Priorat, Spain.
Example Regions	Alsace, France; Priorat, Spain; Douro Valley, Portugal; Swartland, South Africa; Mosel and Ahr, Germany	Mosel, Germany (blue slate); Rheingau, Germany (red slate); Clare Valley, Australia; Priorat, Spain
Grapes That Thrive	Riesling and Pinot Gris (Alsace), Grenache	Riesling, Mencía

Sedimentary Rock

	CHALK	LIMESTONE	SANDSTONE	FLINT/SILEX
Description	This is a specific type of sedimentary rock. You can actually use it to play tic-tac-toe on an asphalt road. Chalk is porous and nutrient-poor and encourages vines to produce highly acidic sap and grape juice. Chalk soil also offers great drainage so the roots don't sit in water and lead to diluted grapes and boring wine.	This type of sedimentary rock is a big, complex family made up of calcium carbonate from seashells and animal fossils. Fossils of everything from tiny oysters to giant whales are constantly adding to the calcium skeletal debris on the ocean floor. Limestone adds a lightness, brightness, and structure to wines lucky enough to be grown on this soil, plus layers of complexity and the sense you are licking a piece of rock, which I happen to love.	This is made from sand particles that have been smooshed under the earth's pressure for millions of years. Grapes grown on sandstone tend to become especially sweet and juicy. (While not completely analogous, this is also true of carrots—if you are produce shopping in France, carottes des sables is a special thing, where carrots still have sand on them to show you they were truly grown in sandy soil.)	This is a great rock for capturing and retaining heat. Flint (also known as silex) can give ripeness in grape-growing regions that might otherwise be too cold for such growing because the rocks absorb heat and then release it back to the plants at night.
Example Regions	Champagne, France	Burgundy, Champagne, Chablis, the Loire Valley, Southern Rhône, Saint-Émilion, and parts of Alsace, France; Rheinhessen, Germany; Sussex, England; Barolo, Barbaresco, and Tuscany, Italy (where limestone is called pietra alberese); California	La Rioja, Spain; Barossa Valley, Australia	Sancerre and Pouilly-Fumé, in France's Loire Valley
Grapes That Thrive	Chardonnay, Pinot Meunier	Chardonnay, Pinot Noir, Nebbiolo, Riesling, Savagnin, Poulsard, Pinot Meunier	Tempranillo, Shiraz	Sauvignon Blanc

ZOOMING OUT: CLIMATE

Climate is so vast a concept it's usually chunked into three more specific terms: macroclimate (big), mesoclimate (medium), and microclimate (small). Macroclimate applies to a large area or a region—like Champagne (France) or Napa Valley (California)—or areas that share a common geography. Any given macroclimate will contain several mesoclimates and countless microclimates. Think of a mesoclimate as more like a specific vineyard and microclimate as a tiny portion of that vineyard, a row or two of vines.

Wine regions can be ascribed to one of three macroclimates: continental, Mediterranean, and maritime. Landlocked regions are known as continental climates and are notably distant from large bodies of water. Continental climates benefit from diurnal shifts—temperature swings between warm days and cool nights, even in summer. Burgundy in France and Columbia Valley in Washington are two examples. Expect a balance of fruit and freshness in wines from these regions.

Mediterranean climates are those named for and located around the Mediterranean Sea, from Spain and North Africa across to Turkey and western Syria. The term also applies to regions that have similar climates, characterized by long growing seasons and moderate temperatures. Southern Oregon, most of Portugal, and Chile's Central Valley are good examples of Mediterranean climates.

Maritime climates are those influenced by large bodies of water. Growing seasons are long, and the climate gets clammy. Vines battle excessive rain and humidity (often leading to various diseases like mold and mildew) throughout the year. Bordeaux, New York's Long Island, and Washington State's Puget Sound are a few examples. In maritime climates, temperature swings are rarer and acidity in the grapes is generally milder.

THINK LIKE A MONK

The Romans officially brought viticulture to modern-day France (or Gaul, as it was known back then), but the Catholic Church most influenced viticulture in the area. This trend began when Charlemagne was crowned Emperor of the Holy Roman Empire in 800 CE. He gave large swaths of land to the Church, cementing its role as the most important landowner in the region. Dukes and nobility followed suit and quickly became a powerful part of both the political and viticultural ecosystems (but more on them later).

One particular order of Cistercian monks split off from the others and built the abbey of Clos de Vougeot. Then they began cultivating its surrounding vineyard, in the twelfth century. These guys deserve special recognition for their contributions to our current understanding of vineyard hierarchy. They painstakingly analyzed and recorded best practices and ultimately set the foundation for Burgundy's *climats* today.

FARMING PRACTICES THAT KEEP SOIL HEALTHY

If you have the chance to visit a vineyard, look between the vines at what's growing. Healthy soil is vibrant, with all sorts of green things like leaves, shoots, and flowers pushing up through the earth. Dead soil is gray, crumbly, and looks like the surface of the moon. This has to do with how the vineyards are farmed, and there are a few terms that are useful to know when you want to order a bottle of something at a wine store and ensure it was not doused with Roundup (see the sidebar on the opposite page).

Some farming terms and what they mean are listed below. You can also ask for bottle recommendations based on farming practices at your favorite wine shop or restaurants.

CONVENTIONAL FARMING: "Conventional" means vines were farmed with synthetic chemicals and fertilizers in an attempt to maximize yields. Farming this way was not actually "conventional" until after World War II. Chemicals include pesticides, herbicides, and fungicides, as well as genetically modified variations of plants (GMO grapevines are still rare—hopefully that doesn't change!). Conventional farming reduces labor cost, which is one reason so many farms opt for it. But it destroys biodiversity and poisons the environment. Roundup, the weed killer (also known as the chemical glyphosate), is the main culprit and is being linked to everything from a rise in cancer to autism, infertility, and general inflammation in the United States and around the world. It is especially toxic for the vineyard workers tasked with applying it.

LUTTE RAISONNÉE: This French term essentially means "we only spray if we absolutely must, and in that case only exactly what we need to, and as little as possible." I like to think of this philosophy as akin to the person who eats organic fruits and vegetables, meditates regularly, and prefers herbal medicines but is happy to take an Advil if she gets a bad headache or the occasional antibiotics when stricken with something like a UTI.

ORGANIC FARMING: According to the United States Environmental Protection Agency, organic farming is defined as "food grown and processed using no synthetic fertilizers or pesticides." This means that the winery is not spraying weed killers, herbicides, fungicides, or pesticides on the vineyards. While it may seem like a no-brainer choice not to farm with poison, there are reasons why wineries choose conventional farming. A main factor is cost. Winemaker Diana Snowden Seysses—who makes wine in Burgundy as well as at her family's Napa estate, Snowden Vineyards—shared that it took her several years to convince her family members they should be farming organically because it costs up to *forty times more* per year.

What Is Roundup (Glyphosate) and Why Should I Care?

We can thank the twentieth century for the insecticides and herbicides used today. World War II produced the insecticides DDT and DFDT, invented by American and German scientists, respectively. Agrochemical corporation Monsanto's contributions to the Vietnam War included the poisonous herbicide glyphosate, which found a new market for homeowners as Roundup, still widely used in many countries—including rampantly in the United States—today.

Roundup, or the chemical glyphosate, is a neurotoxin and a carcinogen used as a weed killer (herbicide) in vineyards and on commercial farms, as well as on golf courses and lawns throughout the United States, China, Canada, and other parts of the world. It was created by Monsanto as a variant of Agent Orange, a poison created to strip trees and plants of their leaves in the Vietnam War (so that people couldn't hide beneath them). When the war ended and the US military no longer needed Agent Orange, Monsanto looked to find new customers. They began marketing it to commercial farms in the late 1970s and, starting in 1990, via television commercials to the US male consumer as a catch-all weed killer solution for family lawns, even purchasing commercial spots during Super Bowls. (The domestic-use version featured a man able to check weed-killing off the honey-do list, having more time to play basketball with his son.) Roundup is water-soluble, which makes it particularly scary; it has been linked to everything from cancer to infertility in humans and is widely recognized as devastating bee communities worldwide. Italy banned the use of Roundup in 2016 and most of the rest of Europe followed suit in 2020. Thanks to recent court decisions against Monsanto, the chemical behemoth (now owned by Bayer) is responsible for $11 billion in repayments to resolve more than 100,000 Roundup settlements. This has not stopped wineries from using Roundup, but hopefully it's a start. You can avoid Roundup by purchasing organic produce and looking for wines labeled as "Made with Organic Grapes" or any bottles carrying Demeter or biodynamic certification.

BIODYNAMIC FARMING: This is a holistic and, arguably, spiritual approach to soil health and farming that follows the lunar calendar. It was originally developed by an Austrian named Rudolf Steiner in 1924. The Biodynamic Association defines biodynamics as "a holistic, ecological, and ethical approach to farming, gardening, food, and nutrition." Soil preparations are made each year to increase microbes and soil health, and during the growing season, vines are sprayed as needed with tisanes, or teas made from organic matter or pulverized minerals, to manage the health and life cycle of the vine.

BEYOND BIODYNAMIC FARMING: Some of the most thoughtful winemakers today study their land and have a reverence and appreciation for their vineyards and the environment at large, as well as a boundless curiosity and openness to experimentation. The first time I visited Domaine Jacques Selosse, in 2012, Anselme Selosse explained that he no longer subscribed to biodynamic farming because he felt it was too dogmatic and not tailored enough for his vineyards in Champagne. (Rudolf Steiner lived and worked in Austria, not Champagne, after all.) Marta Rinaldi in Barolo is adamant about farming and harvesting for the Giuseppe Rinaldi winery based on her intuition. Rajat Parr, now making wines at Phelan Farm in San Luis Obispo, California, has taken inspiration from biodynamic practices but tailored them to his vineyards, working with seawater and kelp, ingredients hard to come by in Steiner's Austria but better suited to coastal slopes. These are just a few examples of winemakers who follow their intuition and curiosity in search of creating great wines.

Austrian philosopher Rudolf Steiner, who lived during the late nineteenth and early twentieth centuries, officially proposed his ideas on biodynamic farming during a lecture series, *Spiritual Foundations for the Renewal of Agriculture*, which he delivered in Poland in 1924. Those lectures formed the basis of biodynamic farming today. Steiner believed that Earth is a living being and that all life-forms are connected. He paid particular attention to the soil, which, in biodynamic farming, is enlivened through several different preparations made to enhance compost, strengthen vines, help with ripening, increase photosynthesis, and ensure vineyard health and immunity.

HEALTHY SOIL IS MAGICAL STUFF

The best vine growers and winemakers right now understand that healthy soil is critically important to making delicious wine. Soil influences everything from a vine's vigor to the way we experience a wine's taste in our glass. Dominique Lafon, vigneron and proprietor of Domaine des Comtes Lafon in Meursault, was one of the first in Burgundy to convert his vineyards to biodynamic farming, which encourages microbial life in soil as opposed to killing it, in the mid-1990s. I asked him his thoughts on the role of microbes in terroir, and he shared that microbes are able to digest the different characteristics of terroir and transmit it to the plants. In short, they're incredibly important!

University of Chicago microbiologist Jack Gilbert concluded an experiment with similar findings in 2015. "Different areas of the plant have different microbiomes (communities of microorganisms living together), but in nearly all cases, the microbiome originates with the soil," Gilbert said in his report, summarized by Matt Wood in an article for the University of Chicago. "Everyone latches on to terroir because that's the most interesting thing about vines, but there's a much larger story here with regards to the role the soil plays in the development of the productivity, disease tolerance and stress tolerance of the plant itself." Microbes in the soil are in constant collaboration with soil, and with the grapevine itself.

Healthy soil is the basis for everything. It provides the foundation from which all other life grows. Healthy soil is a mix of rock and organic matter and is teeming with microbial life, in constant communication with the roots. One tablespoon of healthy soil can have *fifty billion* microbes! Soil like this is an incredible carbon sink, absorbing more carbon than all other plants on earth combined.

97

Biodynamic Farming

Q&A with Diana Snowden Seysses

To better understand how biodynamic farming works in real life, I spoke with oenologist (winemaker) and friend Diana Snowden Seysses. Diana is a chemist by training and the oenologist at Domaine Dujac in Burgundy, where vines have been farmed biodynamically for over a decade. A Napa Valley native, she also helps her family make wine at Snowden Vineyards in St. Helena and, more recently, with her sister and cousins at the winery Snowden Cousins. She is also an outspoken advocate for the need to act quickly on climate change. Below is a summary of our conversation.

Q: How do biodynamics work?

A: With biodynamics, there is a holistic approach that involves attention and decision-making in the soil that is driven by something more macro—the planets in our solar system.

Q: Have you experimented with comparing organic and biodynamic farming?

A: At Domaine Dujac, we decided to approach farming a new vineyard parcel half biodynamically and half organically. The [results] were generally similar but with small perceptible differences. The biodynamically farmed parcel had a more organized canopy—and there was also a little more energy, which a laboratory analysis also confirmed.

Q: How so?

A: With sugar! There is more sugar [and thus more energy] in the grapes farmed biodynamically.

Q: Are there levels of interpretation allowed when it comes to claiming one's vines are farmed biodynamically? How rigidly must one follow Steiner's techniques?

A: There is no one recipe to follow when farming biodynamically. Every vineyard will have its own needs at a given time. That said, there are certain things—including spraying horn manure and horn silica—that are universal to anyone practicing biodynamics. (Horn manure is made from cow manure buried underground in a cow horn during the winter months. When sprayed on the vineyards, it regulates the soil's pH balance and stimulates soil microbial activity, root development, and seed germination. Horn silica, by contrast, is sprayed directly onto the vines and leaves during their growth phases. Called the "spray of light," horn silica promotes vigor and reduces a plant's susceptibility to disease.)

Q: Can you share any observations about how the soil, vines, fruit, and vineyard change when farming is converted from conventional to biodynamic?

A: Once you've given up chemicals and are farming organically, the vines are healthier. The transition from conventional farming to organic farming is the greatest shift. However, the biodynamically farmed vines are more organized, slightly more productive, and they have more energy in the form of sugar.

Roots of the grapevines actually exchange information with the soil in which they grow. The plant asks the microbes in the soil for what it needs, and there is an exchange that culminates in riper grapes with higher concentrations of sugar.

Q: Do you see biodynamic farming practices play out differently in different regions (Napa, Burgundy)? If so, how?

A: Yes, for sure. Everyone does it differently. Ideally you will have learned by doing it with someone on their property. For example, Domaine Dujac's foreman was in a group that got together regularly and had a great deal of practical application before beginning to transition Dujac's vines. Different domaines adjust based on their needs, though. I have certainly seen differences in application among domaines within Burgundy as well as differences between farming practices in Burgundy versus California.

Q: What are farming practices like at Snowden in Napa Valley?

A: At Snowden, it took me several years to convince my other family members we should be farming organically. From a financial perspective, they asked why we needed to start spending thirty-two thousand dollars per year on organics instead of the eight hundred dollars they previously spent on Roundup for the entire property. Since 2016, there are no more chemicals.

Q: Why did you convert? What have you noticed?

A: I was looking for an aliveness. Something you can feel with your body. The point is picking grapes that are alive—and there is just one more extra dimension (beyond organic farming) with biodynamics.

99

In wine, healthy soil is also your passport to a faraway place, plopping in a bit of wherever that wine came from right into your glass. Of course, pieces of soil don't literally end up in your wine. But the soil of any given vineyard is unique, like a fingerprint. Microbes and minerals in the soil interact with water that is eventually absorbed through the roots of the vines and into the grapes, and those grapes become wine that tastes of precisely where it grew. All anyone skeptical of terroir has to do is set up a taste comparison among three bottles of Chardonnay grown in Chablis, Meursault, and Napa Valley, respectively, to know that a lot of magic and distinction in a bottle can be attributed to the soil beneath the vines.

Why and how do rock and soil play such profound roles in a wine? The chemical and mineral composition of a rock determines the natural supply of nutrients it releases into the water supply with which it constantly interacts. Soils contain different minerals, each of which has different properties. Some are better matches for various grapevines than others. Different types of clay are composed of different minerals, which interact with microbiomes differently and have different capacities for water and mineral exchange. We're just beginning to understand the resulting chemical interaction, but it's one that invariably impacts taste and texture in wine.

GETTING DIRTY: SOIL TYPES AND HOW THEY IMPACT THE TASTE OF YOUR WINE

Below are some soils that are likely to come up in conversation if you fall in love with a wine and then start asking lots of questions. The list is in alphabetical order—and it's certainly not exhaustive! While most of the soils listed are general (like clay and sand), I've also called out a few that are pretty special and are only found in specific regions.

CLAY

Merriam-Webster defines clay as "a soil that contains a high percentage of fine particles and colloidal substance and becomes sticky when wet." Clay, in the wine-growing sense, has wonderful water-retention properties. In terms of how grapes grown on clay impact wine, think of clay as a muscle—wines grown on clay tend to be rich and concentrated, bold, and quite powerful.

REGIONS: Pomerol, France; Barossa Valley, Australia

GRAPES THAT THRIVE: Merlot and Sangiovese

GRAVEL

This term can refer to stones as small as pebbles and as large as golf balls. Like sand, gravel can be made from any type of rock. What's most relevant is gravel's ability to absorb and refract light and heat, enabling vines in historically cooler climates, like Bordeaux, to ripen grapes when they wouldn't otherwise. Châteauneuf-du-Pape's galets (giant pebbles, roughly the size of a chicken egg) are another famous example, keeping grapevines warm throughout cool nights. Gravel also has a direct impact on a wine's taste. I have not gotten a straight answer as to why wines grown on gravel taste, well, gravelly, but I surmise it has something to do with more surface area of those tiny rocks interacting more with the water in the vineyards and the microbes in the soil working their magic, resulting in a more perceptible minerally taste.

REGIONS: Médoc and Graves (Bordeaux, France); Châteauneuf-du-Pape (Rhône, France); Gimblett Gravels (New Zealand); Walla Walla, Washington

GRAPES THAT THRIVE: Grenache, Syrah, Cabernet Sauvignon, Merlot

LOAM

Loam is an incredibly fertile combination of sand, silt, and clay. It usually results in vines that make large quantities of juicy but otherwise unremarkable grapes, unless loam is combined with other soils, like sand, which gives the wines more flavor. Much of the "valley" portions of Napa and Sonoma are of sandy-loam soils. Grapes grown on them make lush, juicy, and opulent wines.

REGIONS: Napa Valley and Sonoma Valley, California

GRAPES THAT THRIVE: Cabernet Sauvignon, Merlot, Pinot Noir (expect juicy, fruit-driven examples of these grapes from sandy-loam soil)

MARL

This crumbly stone made from a mix of limestone and clay is commonly found in France and northern Italy. Marl comes in different colors (blue, gray) and is the bedrock for some of the greatest vineyards in the world. Marl, like any limestone, adds an elegant structure and backbone to the wine, as well as the perception of brightness and acidity. For me, it adds another two or three dimensions to a wine, also increasing its length on the palate and adding general complexity. This phenomenon is not scientifically proven, it's just something I continue to find true over and over again.

REGIONS: Parts of Burgundy and the Jura, France

GRAPES THAT THRIVE: Pinot Noir, Poulsard, Trousseau, Chardonnay, Savagnin

SAND

To quote author Alice Feiring in *The Dirty Guide to Wine*, "Sand is a texture, not a rock. It can be made up of anything that is pulverized." Sand has terrific drainage and tends to result in juicy, fruit-forward wines. As an added benefit, phylloxera cannot live in sandy soil, so regions and vineyards with these soils can grow ungrafted vines. I saw this for the first time while working a harvest at Bodega Chacra in Río Negro, Patagonia. Unlike most regions, which get their vines from a nursery, Bodega Chacra can plant its vine cuttings by splicing off part of the vines they want to clone and planting them in a big open-air sandbox.

REGIONS: Barossa Valley, Australia; Lodi, California; Colares, Portugal; parts of Châteauneuf-du-Pape (Rhône, France); northern Médoc and Graves (Bordeaux, France); Río Negro (Patagonia, Argentina)

GRAPES THAT THRIVE: Cabernet Sauvignon, Zinfandel, Grenache, Pinot Noir

REGION-SPECIFIC SOILS

BLUE MARL: Marl is stone formed from limestone-rich mud. Blue marl is a type of limestone found in the Jura region. Its color is a pale blue, like an Araucana chicken's egg, and it produces structured, long-aging wines generally from Savagnin or Chardonnay that taste, for lack of a better description, extra limestone-y.

JORY: This is a red, clay-loam volcanic soil with a high iron content found in the hills surrounding the Willamette Valley in Oregon. It has been recognized as Oregon's official state soil since 2011. In my experience, wines grown on this soil exhibit a warmth and a roundness, and Pinot Noirs grown on Jory soil remind me of squash blossoms (shout-out to Bobby Stuckey who first mentioned this to me—now I smell it every time).

KIMMERIDGIAN LIMESTONE: This famous limestone, also known as *terres blanches* ("white soils"), gets its name from the town of Kimmeridge in southern England, and it runs down through some of the most famous wine regions in the world, from Champagne to Sancerre and Chablis. This limestone specifically refers to a soil formed between 152 and 157 million years ago and contains fossilized baby oyster shells in the shape of commas and of the species *Exogyra virgula*. They get their name from their shape (*virgule* is the French word for "comma"). What is most special about this soil, beyond that it produces remarkably age-worthy wines, is that drinking wine from grapes grown on Kimmeridgian marl tastes like you are drinking the sea.

PONCA: This is a mix of marls and sandstone, plus marine fossils from what is now the Adriatic Sea, brought to the earth's surface sixty-five million years when tectonic plates shifted, Africa and Eurasia collided, and the Alps shot up into the sky. The prized soil is found in the Colli Orientali region of Friuli-Venezia Giulia, Italy, and across the border in Slovenia. Ponca soil helps to give wines strong mineral notes and a remarkably age-worthy quality.

Stellenbosch
PROVENCE
Minho
Piedmont
Jura
COQUIMBA
RIOJA
MITTELRHEIN
MOSEL
SWAARTLAND
Puglia
WACHAU
SALTA
LOIRE
Wine list

GRAPES YOU'LL LIKELY ENCOUNTER

Up to this point we've been focused mainly on details that provide nuanced but vital imprints on a wine—things like soil, farming practices, winery equipment, and a winemaker's point of view. These are all important factors—but nothing plays a bigger role in how a wine tastes than the grapes themselves. This chapter is here to provide an overview of grapes you are going to encounter most often and what to expect from them. Yes, we'll cover Chardonnay and Cabernet Sauvignon, but hopefully you'll add some new favorites to your list, or at least get a better sense of why you like them!

There are approximately *ten thousand* types of known grapes in the world, and if you want to learn more about them, there are amazing books dedicated to exactly that! You will not see obscure grapes like Gringet or Blauer Wildbacher here, because only a small handful of restaurants or wine shops will have taken the time necessary to understand the grape, source the best examples, and be able to defend their placement on a list. Think of this section as a starting point. Finally, I've divided the list into white grapes and red grapes and have ordered each alphabetically.

A NOTE ON ACIDITY: In the snapshots beneath each grape description, the acidity line references total acidity, also known as TA, which encompasses the sum of a wine's organic acids—tartaric, malic, citric, and lactic. Acidity doesn't exist in a vacuum—a high-acid wine, like German Riesling, often tastes much less tart when accompanied by some residual sugar. (Think of lemonade—it's bright and balanced, compared to mouth-puckering lemon juice, which has the same amount of acidity.) Lots of other factors can influence acidity in a wine, ranging from how a vine is pruned, how early the winemaker harvests, and whether or not the winery is choosing to "acid-correct" during fermentation. In wine, everything is contextual. That's why the descriptors in the pages that follow are presented in a range and are meant to serve as a guideline, not a holy grail. As a reference, TA in wine generally falls between 4.5 grams per liter (the minimum amount of TA allowed in wines from the European Union) and 9 grams per liter (g/l).

A NOTE ON PH: It's worth noting that acidity is also a function of a wine's pH, a term that gets thrown around a lot in wine circles. A frank (if somewhat clunky) definition comes from the United States Department of the Interior, which explains pH as "a measure of how acidic/basic water is. The range goes from 0 [to] 14, with 7 being neutral. pHs of less than 7 indicate acidity, whereas a pH of greater than 7 indicates a base." White wines fall in a range of 3.1 to 3.5 pH whereas red wines fall in a range of about 3.5 to 4.2 pH. While pH can certainly influence the way acids taste, I've chosen to focus on the actual acidity ranges, as those are stable throughout the fermentation process whereas pH is influenced by a variety of factors.

A History of Noble Grapes

The term *noble grape* is generally and historically understood to reference varieties considered most capable of creating the highest-quality wine when all other factors are equal. I took this term for granted until writing this book, which prompted some digging into its historical origins. Turns out the origins are French (the original term was *cépages nobles*) and popularized in Britain (which still has a monarchy). Six grapes were historically "noble," five of which have French roots (Chardonnay, Sauvignon Blanc, Merlot, Pinot Noir, Cabernet Sauvignon). The sixth (Riesling) is a German variety, also grown in the French region of Alsace. Speaking of Alsace, this region has its own *separate* set of four noble grapes—Riesling, Gewürztraminer, Pinot Gris, and Muscat—all codified into Alsatian wine law. These grapes are recognized as the top performers by the Alsatian powers that be, and they are the grapes used for any Grand Cru vineyard bottling with a few exceptions—a vineyard named Zotzenberg, which can also include Sylvaner, and vineyards Vorbourg, Kirchberg de Barr, and Hengst, which can include Pinot Noir.

The term "noble grape" also carried weight during the early days of the American wine scene. In his seminal book, *Frank Schoonmaker's Encyclopedia of Wine*, American travel guide and wine writer Schoonmaker wrote: "A noble variety is one capable of giving outstanding wine under proper conditions, and better-than-average wine wherever planted, within reason," and he goes on to define "noble wine" as "one that will be recognized as remarkable, even by a novice."

Today, eighteen grapes benefit from the "noble" endorsement (the additions include the red grapes Grenache, Malbec, Sangiovese, Tempranillo, Syrah, and Nebbiolo—and the white grapes Sémillon, Viognier, Pinot Grigio, Chenin Blanc, Muscat, and Gewürztraminer), though it's fair to say the world of wine is quickly outgrowing this sort of categorization. There is no one definitive group of grapes deemed certain of producing greatness. I've had my share of terrible wine from noble grapes—and I've had incredible wines from "humble" grapes like Listán Negro when grown and vinified in capable hands in the Canary Islands.

Plus, a lot of what defines excellence is contextual. If you're eating a Margherita pizza, the best option might be a bottle of Barbera d'Alba, even though the noble grape in the Piedmontese neighborhood is Nebbiolo (which is high in tannins, acidity, and alcohol), and probably a bit much for that pizza unless you've gone out of your way to find a very elegant bottle with age from a producer in a delicate vintage, and your pizza is covered in meat. As Eric Asimov wrote in his *New York Times* article on precisely this subject of noble varieties, "[a] caste system for grapes is a backward-looking approach to a world with wonderful possibilities ahead of it."

GRAPES YOU'LL LIKELY ENCOUNTER

A NOTE ON TANNINS: White wines have very little to no tannins, so you won't see tannins for white grapes listed in the pages that follow. Most tannins are extracted during maceration, a process largely reserved for red and orange wines. I've included a range for tannins in the red wine descriptions, though, because different types of grapes tend to skew more tannic than others. As with anything, these ranges are relative. Any winemaker anywhere is capable of creating an atypical wine with acrobatics in the winery. The ranges in this book spotlight typicity, not outliers.

A NOTE ON BODY AND MOUTHFEEL: *Body* refers to the texture and weight of a wine while *mouthfeel* is more about the sensations it creates on your palate. Tannins, acidity, and alcohol all play a role in mouthfeel. A wine high in tannins will leave your mouth feeling quite dry. High-acid wines will leave your palate puckering, while low-acid wines will not.

When it comes to explaining a wine's body, the best analogy I've heard involves milk. A full-bodied wine is akin to whole milk or even half-and-half (in the case of very high alcohol wines, like Barossa Valley Shiraz), whereas medium-bodied wines can range in palate weight from 1 percent to 2 percent milk, and a light-bodied wine is more similar to skim milk. If you are vegan or lactose intolerant and are displeased with that analogy, consider a full-bodied wine on par with macadamia nut milk, light-bodied wine comparable to oat milk made without guar gum or stabilizer, and a medium-bodied wine in between the two.

Typically, a wine's alcohol content is proportional to its body. Wines high in alcohol tend to be full-bodied, because alcohol carries palate weight thanks to glycerol, a byproduct of fermentation described as a "colorless, odorless, sweet-tasting syrupy liquid" by Sciencedirect.com. A winemaker's decisions during vinification and élevage influence body, too. For example, choosing to leave the wine in contact with lees will add more weight and richness to the wine. And choosing oak or a porous vessel (as opposed to stainless steel or something non-porous) will add richness and texture as well (see page 77).

A NOTE ON FINISH OR COMPLEXITY: I haven't included finish or complexity below—but please note any producers I mention in this book are ones I think are the very best examples! Finish, or complexity, is going to be producer-specific and not predictable based on grape variety alone. Wines from the same vineyard, vintage, and grape can bear no resemblance in complexity if one producer prioritizes quality and the other takes shortcuts. (See page 68 for an example of wineries making crappy wine with expensive vineyards.)

WHITE GRAPES

Technically, these are green, yellowish green, golden, or occasionally pinkish in color, but the grapes below are those commonly used to make what we call white wines, listed alphabetically.

ALBARIÑO

Albariño, a thick-skinned, aromatic, and delightful grape, has captured the world's imagination and is an elevated alternative to the sea of overcropped Pinot Grigio from Veneto for anyone in search of a crisp, bright white. Native to Galicia (northwestern Spain) and its neighbor Portugal (where the grape is called Alvarinho), it hits all the right notes—bright acidity, unflappable body, juicy and intriguing flavors (tangerine, green apple, apricots, and honeysuckle) with a minerally backbone. It comes in oaked and unoaked versions and is quite good both ways.

Today, Albariño is grown everywhere from California, Oregon, and Washington State to New Zealand and Australia. It's also showing up in places as unlikely as Uruguay and, since 2010, is even allowed in France. The grape's thick skins help protect it against rot in humid climates.

If you want to create your own little heaven, pair a good bottle with pan con tomate or, if you're feeling indulgent, a slice of sourdough bread slathered with butter and accented with two or three cured anchovies and some flat parsley leaves. Albariño with a grilled turbot, poached cod, or roasted halibut is great, too, and if you're not particularly hungry, it's equally good with a view of your backyard.

ALBARIÑO SNAPSHOT

COLOR: pale yellow with hints of green (and gold hints for wines aged in oak)

ACIDITY: medium+ TA

ALCOHOL: 11.5 to 12.5 percent ABV (as low as 10.5 percent if you're drinking it as Vinho Verde from Portugal)

BODY AND MOUTHFEEL: medium to medium+ (depends on whether or not it was oak-aged but either way, it has good structure thanks to those thick skins)

FLAVORS AND AROMATICS: Meyer lemon, tangerine, green apple, apricot, nectarine, verbena, honeysuckle, ocean spray, limestone

ALIGOTÉ

Best known as "Burgundy's other white grape," Aligoté is a sibling of Chardonnay without the weight, the historic nobility, or the price tag. Aubert de Villaine, proprietor of the world-renowned winery Domaine de la Romanée-Conti, is so delighted by Aligoté that he and his wife, Pamela, created a winery in Bouzeron, a small appellation in the southern part of Burgundy dedicated exclusively to it. Domaine de Villaine makes delicious and well-distributed wines that retail for about forty dollars per bottle if you are keen to try them yourself. Historically Aligoté was the preferred wine for an aperitif called a kir (white wine with a dash of blackcurrant liqueur—not to be confused with a Kir Royale, blackcurrant liqueur with Champagne).

Aligoté, at its best, is zippy and bright, with notes of fresh red and yellow apple, Meyer

109

lemon, and limestone. It is an acceptable stand-in for Chardonnay when you're feeling exploratory and want to drink something that's neutral, delicious, and versatile enough to match with anything—from a slice of Comté cheese to a pan-fried sole fillet to a warm salad of roasted squash.

CHARDONNAY

Chardonnay, one of the world's most famous white grapes (originally from Burgundy but now grown everywhere from Australia and the United States to Chile, Serbia, and Ukraine), is inherently neutral. It had a low point in the 1990s for tasting like liquefied creamed corn, but Chardonnay is really more like a white canvas whose taste reflects where it was grown and who made it. My mom is among the many people who thinks she does not like Chardonnay while simultaneously loving wines from Chablis—the northernmost region in Burgundy (which produces wines exclusively made from the grape). She is still recovering from a prior era of drinking cheaply made bottles flavored with "oak essence" with her best friend Debbie on our back porch.

Even within Burgundy, Chardonnay expresses itself very differently throughout the subregions, though Chablis is so distinct it's worth spotlighting here. (We'll cover Burgundy more in chapter 6, page 142.) In Chablis, Kimmeridgian marl works its magic, turning Chardonnay into an elixir of seashells, ocean mist, fresh lemon, white chocolate, and fresh mint. Historically Chablis was not as wealthy as its southerly counterparts in the Côte de Beaune, so new oak was a rarity—and not part of its stylistic identity in the way that it is farther south.

Some of my favorite Chardonnays grown outside of France are from the United States (especially California's cooler Santa Barbara County) and Western Australia, where brisk ocean breezes give the wines a Key-lime-zest acidity and lots of balance. Napa Valley makes some very famous, very opulent versions that rely heavily on new oak and taste almost tropical, like mango and marzipan. This style is not my thing, but plenty of people love it. Part of Chardonnay's charm is its global popularity—there is so much to explore, and any attempt to shoehorn its identity would be frankly insulting to this extraordinary grape. There's a Chardonnay for any occasion. Bourgogne Blanc or Santa Barbara Chardonnay is terrific with a summer dish of corn, clams, and Sungold tomatoes—as well as something less specific like roasted fish or even spaghetti alle vongole. Richer examples are great with roast chicken. Go forth and explore; there's a whole world waiting.

ACIDITY: medium to medium+ TA

ALCOHOL: anywhere from 11.5 to 15 percent ABV, depending on where it's grown

BODY AND MOUTHFEEL: medium+ (like 2 percent milk)

FLAVORS AND AROMATICS: fresh apples (green, gold, and red), sunflowers, fresh lemon, Meyer lemon, lemon curd, yogurt, wet stones, hazelnuts, almonds, mango, marzipan; these vary enormously based on region

CHENIN BLANC

Chenin Blanc is easy to love because it is zippy and bright, but with presence and texture, and because it's slightly aromatic—like green and red apples with lemon and lavender honey. It also transmits a sense of place so well. With roots in France's Loire Valley, it is now also grown throughout South Africa (where it's called Steen), Australia, the United States, and nearly everywhere that's making great wine.

In Chenin Blanc's birthplace, Anjou and Saumur in the Loire Valley, the grape grows on wildly different soils according to geography. Soils in the western half of Anjou are known as L'Anjou Noir, named for the dark color of rocks found there—schist, slate and spilite, among others. Soils in the eastern half are pale and creamy in color, with an abundance of limestone and marl. Vouvray is famous for tuffeau, its porous limestone that is so soft and crumbly that big tunnels and wine cellars have been carved underground throughout the region, and the wines taste of chalk and mineral, too.

The grape itself tends to have oxidative qualities (like a red apple, sliced and left on the counter for an hour or two) when it's picked late or the grapes are botrytized, and can also be bright and citrusy, with fresh green apple notes and lemon zest.

A refreshing sparkling Vouvray (a minerally and bright wine from the Loire Valley) is a slam dunk with a crudités platter or a simple crudo dish, while a still wine from Savennières, about 100 miles east along the Loire River on decomposing schist soil, is bold and powerful and great with anything from grilled sausages to a whole-roasted fish. Chenin Blanc is also behind many a great dessert wine, where it's a delight with apple pie or a fruit tart.

CHENIN BLANC SNAPSHOT

ACIDITY: high TA

ALCOHOL: 11 to 13 percent ABV

BODY AND MOUTHFEEL: medium to full (I know this is an absurdly large range, but that's because there are so many different expressions of Chenin and these factors really depend on where the grapes are grown and how they're made into wine. Chenin Blanc is a chameleon and can be bright and refreshing when grown in a cool climate, or it can be rich and luxurious in a sun-drenched appellation like Savennières in the Loire Valley, let alone elsewhere in France or even other countries. No matter what, expect plenty of refreshing, often mouth-puckering acidity from this glorious grape.)

FLAVORS AND AROMATICS: lemon, bruised red and green apples, white chocolate, cheese rind, quince, honeysuckle

GRÜNER VELTLINER

Native to Austria, Grüner Veltliner is the country's most planted grape and covers approximately one-third of its vineyard land. The world's best examples are Austrian, although some very good Grüner Veltliner is now also being grown elsewhere, like Sta. Rita Hills in California. Grüner Veltliner tastes like fresh green vegetables sprinkled with cracked white peppercorn. It's a terrific pairing with crisp salads or green vegetables. It's also a rare match with asparagus, a notoriously finicky vegetable when paired with most wines. Most of my favorite expressions of Grüner Veltliner are from the Wachau region in Austria, where the vines benefit from cool temperatures and steep, rocky slopes, and the grapes are ripened by the Danube River, which reflects light onto the vines like a big mirror. Humid conditions in some vineyards are also perfect for botrytis (see page 54), which can add another level of complexity and more flavor to the wines. Botrytized wines are not necessarily always sweet—think of them as having an extra burst of texture and flavors of ginger, chamomile, and saffron.

GRÜNER VELTLINER SNAPSHOT

ACIDITY: medium+ TA

ALCOHOL: 11.5 to 14 percent ABV, depending on where it's grown and what style the winemaker is making

BODY AND MOUTHFEEL: medium+

FLAVORS AND AROMATICS: red and green apples, asparagus, green lentils, white pepper, lemon, ginger, chamomile, wet stones

MARSANNE / ROUSSANNE

These grapes, which are native to France's Northern Rhône, are all about texture and richness—not acidity, which make them the black sheep of the white wine world. Wines from these Northern Rhône grapes are round and luxurious and taste of almonds, vanilla, and melon. Often aged in oak, they are creamy and soft in texture and various shades of golden in color. Marsanne and Roussanne are the grapes in Hermitage Blanc, which Francophile and founding father Thomas Jefferson called "the greatest wine, without exception." Additionally these grapes are grown in Crozes-Hermitage and Saint-Joseph (both also regions in the Northern Rhône), as well as blended with other varietals in the Southern Rhône, where their identity is less distinct. California and Australia are responsible for many good examples as well.

One of the most formative pairings I've ever had was a bottle of Hermitage Blanc with a bowl of crayfish bisque, served to me by Erin Chave of Domaine Jean-Louis Chave, who lives on the hill of Hermitage in the Northern Rhône. The rich textures of the bisque and the wine played off of each other like poetry, turning texture into the centerpiece of a pairing. These wines are also great with an array of cheeses, roasted meats, fish, and even pasta, if you are not much of a bisque person. An uni pasta, now that I think of it, would be quite good here, too.

MARSANNE / ROUSSANNE SNAPSHOT

ACIDITY: low TA

ALCOHOL: 13 to 15 percent ABV

BODY AND MOUTHFEEL: full, rich, round

FLAVORS AND AROMATICS: raw and toasted almonds, acacia honey, vanilla, wet stones

MELON B

The Muscadet appellation, in France's Loire Valley, is the epicenter for Melon B, formerly known as Melon de Bourgogne. However, winemakers are beginning to grow this grape elsewhere in North America, too (notably Oregon, Washington, and California). As its name suggests, the grape originated in Burgundy and grew there until the powerful oenophile Philip II the Bold, Duke of Burgundy (son of France's King John II—we'll cover the duke more on page 143), ordered its destruction; he deemed Melon less noble (and less profitable) than Chardonnay and supposedly called it and other grapes "harmful to human health." Melon de Bourgogne found its new home in Nantes after an especially harsh winter in 1709, when cold temperatures destroyed all of the region's less durable vines. In response, a septuagenarian King Louis XIV demanded that the growers in Nantes replant their vineyards with hardy Melon de Bourgogne.

Today, Melon de Bourgogne (or just Melon, on labels outside of France) is a popular pairing with oysters thanks to their shared briny salinity. In France, anything labeled as Muscadet is made entirely from this grape. Expressions are a little bit savory and even herbaceous, as if someone plucked a sprig or two of thyme and tossed them into the fermentation vat. Melon reminds me of a very good lemon vinaigrette—citrusy, salty, bright—and is a great match for everything from almonds and olives to fresh shellfish or a Niçoise salad.

MELON DE BOURGOGNE SNAPSHOT

ACIDITY: medium+ TA

ALCOHOL: 11 to 12 percent ABV—this grape is mostly planted in cool regions and its brightness reflects that!

BODY AND MOUTHFEEL: crisp and tart with additional texture a result of time spent aging on the lees

FLAVORS AND AROMATICS: lemon, brioche, green apple, ocean spray, fresh thyme

PINOT GRIGIO / PINOT GRIS

A member of the Pinot family (and the most popular mutation of Pinot Noir), this pink-skinned grape originated in Burgundy, France, then made its way to Italy by way of Piedmont. Italy, where it is called Pinot Grigio, gave the grape its star power.

The wines are best known for being slightly aromatic, usually unoaked, and dry. Thanks to skyrocketing popularity in the 1990s, there is an ocean of charmless and overpriced Pinot Grigio out there. But that is because of commercial winemaking and vineyards that would probably be better served as wheat fields—it's not the grape's fault! Pinot Grigio, which is called Pinot Gris in Alsace, France, and in Oregon, can be divine.

The grape is widely grown throughout Europe and goes by different names in each region. Look for it as Grauburgunder (*gr-OW-burgunder*) in Germany, Malvoisie (*mal-vwah-zee*) in France's Loire Valley and Savoie regions, and Szürkebarát (*ZUR-cab-a-rot*) in Hungary. While Oregon was an early adopter in the "New World," Pinot Gris is now grown everywhere from California and Washington State to Argentina, Chile, Australia, and New Zealand—where it's nearly overtaking the second-most-planted white grape, Chardonnay.

As far as taste goes, Pinot Gris/Pinot Grigio is a hard one to pinpoint unless you generalize by country—or, in France, by region. Pinot Gris in Alsace benefits from sunny days, a mosaic of interesting soils, and a regional history of letting the Alsatian noble grapes (of which Pinot Gris is one) flirt with a bit of botrytis. Additionally, the wines are often aged in large oak barrels, which gives them a round and luxurious texture. Pinot Gris wines from Alsace are almost otherworldly—think apple blossoms, honeysuckle, ginger, chamomile, quince, and candied citrus. If you spot a Pinot Gris made by Domaine Albert Boxler, buy it and drink it with anything from pad Thai to quiche Lorraine. Or, honestly, just by itself. It's so good.

Italian Pinot Grigios are generally aged in steel tanks and are lighter, brighter, and slightly aromatic—think melon, white peach, lemon zest, and peanut shells. The peanut-shell note comes from some additional lees aging—it's one way to add texture to wine that's far less costly than oak. Most of the time I think of Pinot Grigio as more of a snacking wine—drink a glass alongside a bowl of cashews, sliced apples, or a wedge of Appenzeller or another hard alpine cheese. There are exceptional Pinot Grigios out there—Venica & Venica makes one called "Jesera" that has been a favorite and a staple on every list I've ever written since I started buying wine. Drink it with anything.

PINOT GRIGIO / PINOT GRIS SNAPSHOT

COLOR: slightly pink (because the grape skins are bright pink!)

ACIDITY: medium+ TA

ALCOHOL: 11.5 to 14.5 percent ABV, depending on where it's grown

BODY AND MOUTHFEEL: hard to pigeonhole—light and refreshing as Pinot Grigio; luxurious and round when grown in Alsace; moderate in most places the world over

FLAVORS AND AROMATICS: lemon, brioche, green apple, melon, white peach, apricot, ginger, quince, apple blossom, candied citrus

RIESLING

Riesling is everything I dream of in a wine, when grown thoughtfully in low yields on interesting soil in a cool to moderate climate.

While experts believe that the grape originated in Germany's Rhine River Valley, it now thrives in cool climates across the globe. Riesling is on record as being planted as early as 1477 in Alsace, France (its quality was praised by the Duke of Lorraine), and German immigrants brought it to the United States in the early nineteenth century. Around the same time, an Australian botanist named William Macarthur introduced Riesling to New South Wales, where it became the most planted white grape in Australia until the early 1990s, when Chardonnay finally overtook it. Today, people are growing it everywhere, from New Zealand and Chile to New York, Washington State, and California, with experimental wineries planting it throughout the globe. Riesling is high in acidity and aromatics—like lime blossoms, ginger, tangerine peel, and green apples. It's terrific with everything from green salad to charcuterie or dumplings to a bowl of chips.

Many people think of Riesling as a sweet wine—and sure, there are many great examples—but since the 1950s, after the invention of sterile filtration, sweet is a mostly stylistic decision, and many of the greatest examples I've had are dry.

RIESLING SNAPSHOT

ACIDITY: medium+ to high TA

ALCOHOL: 8 to 13 percent ABV, depending on where it's grown, and if it is fermented dry

BODY AND MOUTHFEEL: depends on sweetness—but its finish is always bright and refreshing, thanks to all of that acidity

FLAVORS AND AROMATICS: lime zest, tangerine, green apple, ginger, chamomile, wet stones

SAUVIGNON BLANC

Like so many other international grape varieties, Sauvignon Blanc is native to France and currently found pretty much everywhere grapes are grown. Whether from New Zealand, Chile, the United States, or anywhere else, Sauvignon Blanc has skyrocketed in popularity and is beloved for its bright and refreshing qualities, plus its gooseberry, green pepper, and freshly cut grass aromas. This grape is also a terrific conductor of terroir, picking up flinty notes in Sancerre and riper, tropical fruit in warmer climates like Napa Valley. In New Zealand, it keeps a bright, citrusy focus. In short, there are endless ways to get to know and love Sauvignon Blanc.

The name Sauvignon Blanc is believed to have originated from the French word *sauvage*, or "savage." As legend has it, the variety grew wild in southern France, especially around Bordeaux. In the 1880s, an Australian statesman and explorer, Charles Wetmore, brought the first Sauvignon Blanc cuttings from Château d'Yquem in Bordeaux to California and then planted them in the Livermore Valley, Australia. Centuries later, Robert Mondavi made wines from this grape in California and renamed it Fumé Blanc, believing his revision had a more mellifluous ring and better marketing potential. A fun fact: Sauvignon Blanc, along with Cabernet Franc, is a parent of Cabernet Sauvignon.

Pairing-wise, there are a lot of ways to enjoy these wines—they shine alongside any green cruciferous vegetable, as well as fresh and bright flavors—like a citrusy fluke crudo or anything spotlighting goat cheese. This grape makes wines that are vivacious, confident, aromatic, and the opposite of neutral.

SAUVIGNON BLANC SNAPSHOT

ACIDITY: medium+ TA

ALCOHOL: 12 to 13.5 percent ABV

BODY AND MOUTHFEEL: bright and refreshing

FLAVORS AND AROMATICS: fresh lime, gooseberry, kiwi, passion fruit, green bell peppers, freshly cut grass, tarragon, chives, wet stone, flint

115

GRAPES YOU'LL LIKELY ENCOUNTER

VERMENTINO

Just by drinking a glass of this grape native to Italy, you're transported right over to the Italian coast, something I opt for on the regular. Bright and saline, with notes of citrus and herbs, it's a no-brainer match with fish of any kind—fried, as in fritto misto, in pasta, or cured in salt (anchovies!)—and draped across good bread slathered with butter and parsley. There are so many good ways to enjoy Vermentino that, although you won't see it on every wine list, you'll see it on several. In Liguria, it's also known as Pigato. In Corsica, it's called Vermentinu. In Piedmont, it's called Favorita, and in southern France, it goes by Rolle.

VERMENTINO SNAPSHOT

ACIDITY: medium+ TA

ALCOHOL: 11.5 to 12.5 percent ABV

BODY AND MOUTHFEEL: medium

FLAVORS AND AROMATICS: green apple, lemon, bay leaf, oregano

RED GRAPES

It wasn't until I started to taste wine analytically that I began to understand the range of difference between a thin-skinned, low-tannin grape like Gamay or a thicker-skinned, high-tannin grape like Cabernet Sauvignon. Red wine is a wonderland of possibilities. A few star players are outlined below.

CABERNET FRANC

Cabernet Franc is a delightful grape that, while charming on its own, is usually blended with Cabernet Sauvignon and Merlot (its more famous progeny) in Bordeaux and Bordeaux-inspired blends throughout the world. Native to western France, Cabernet Franc is notably lighter-bodied than Merlot or Cabernet Sauvignon, with mild tannins and a distinct raspberry-and-lead-pencil-shaving profile thanks to its high pyrazine content (see page 34). It's also bottled as a varietal wine or blended with a touch of (no more than 10 percent) Cabernet Sauvignon in the Loire Valley villages of Chinon, Bourgueil, and Saint-Nicolas-de-Bourgueil, all terms you will see on wine labels.

While Cabernet Franc is regarded as a lighter-bodied red, there are exceptions to the rule. Château Cheval Blanc in Saint-Émilion makes an iconic expression of the grape, as does its world-famous neighbor, Château Lafleur in Pomerol. Cabernet Franc ripens earlier than Cabernet Sauvignon, which makes it popular in cooler climates like northern Italy and

Switzerland. As a blending grape, it's planted throughout the globe and adds aromatic complexity to Bordeaux blends everywhere from Napa Valley in California to Margaret River, Australia. Cooler-climate examples are delicious alongside a plate of charcuterie or a roast chicken, while fuller-bodied versions are excellent with racks of lamb, grilled pork chops, portobello mushrooms, or a lean and luxurious filet mignon.

CABERNET SAUVIGNON

This grape, a child of Cabernet Franc and Sauvignon Blanc, has created quite the global footprint, considering it did not even exist until the seventeenth century, thanks to a chance crossing in southern France. Château Mouton Rothschild in Pauillac is the first winery on record to have worked with the grape, and now Cabernet Sauvignon's mark is so vast it spans the entire world of wine, from France to Lebanon, Chile, and Australia—everywhere grapes are grown. Wine writer Jancis Robinson calls it a "colonizer," leading to the uprooting of other indigenous varieties. It currently holds the spot of most-planted wine grape in the world.

While initially made famous in Bordeaux, it is not the most prevalent variety there (Merlot is the most-planted red grape, with Cabernet Franc right behind). Cabernet Sauvignon's global rise can be attributed to a slew of practical factors related to its relative ease in cultivation: it ripens late, avoiding spring frosts; has thick skins that protect against mildew and rot; and holds its typicity regardless of where it's planted. It's also known for high tannins and a full body, qualities that Robert Parker and subsequent influential critics found desirable as they set out to define quality based on their "100 Point" systems, which prized high amounts of alcohol, tannins, and body as arbiters of greatness.

The grape is notable for its especially high tannins and minty, bell-peppery smell, thanks to pyrazines. It's always terrific with a grilled juicy steak, or red meat of any kind, really. And whether you like ripe, jammy wines or more gravelly mineral ones, there is a world of styles available.

GAMAY

Banished from Burgundy in 1395 by Burgundian duke Philip II the Bold (the same guy who ousted Melon B; see page 113) for its "very great and horrible harshness," Gamay waited in the wings for several centuries before finally coming into the limelight in the twenty-first century. (In fairness, Philip probably aimed to protect Burgundy's reputation for quality, and Gamay is a higher-yielding plant, which can result in more dilute wines for growers who aren't quality-oriented.) Philip the Bold relegated Gamay south to Beaujolais, where it flourished on the granite soils. In contrast to Pinot Noir, Gamay ripened earlier, proved easier to grow, produced significantly higher volumes, and made fruitier wines that were hard not to love.

Gamay is a global superstar these days, made famous by ten villages in France known as the Cru and a handful of gifted growers who believed in the grape's potential when others did not. While the clay-based soils in the south, labeled simply "Beaujolais," tend to produce some fairly lackluster wines, the ten Cru in the northern hills of Beaujolais are their own separate entity entirely. Located on an array of granite soils in the northern hills, these villages are so superior to the surrounding generic Beaujolais appellation that they are each labeled separately by village. You can read more about them in the Beaujolais section (see page 148).

Today, Gamay is grown in California, Oregon, and Australia. It has become almost a cliché to suggest pairing Gamay with Thanksgiving, but clichés exist for good reason, and Thanksgiving dinner is, in fact, a very good match. Gamay is also a red wine so low in tannins that you can drink it with something as gentle as oysters. It's also delicious with glazed duck breast, Thai food, and anything with a bit of spice.

GAMAY SNAPSHOT

COLOR: ruby, with a pink rim

ACIDITY: medium to medium+ TA

ALCOHOL: 12 to 13.5 percent ABV

BODY: light and bright

TANNIN: low to medium

FLAVORS AND AROMATICS: raspberry, granite, red peach, violets, potting soil

GRENACHE / GARNACHA (ROUGE)

Grenache's origins are claimed by both Sardinia (where it's called Cannonau) and Spain (where it's known as Garnacha), and, for a time in the late twentieth century, it held the title of second-most-planted grape in the world. It's worth noting that Grenache comes in three forms—rouge (or red), which is its best known, as well as two mutations, gris (gray, or more accurately, pink) and blanc (white). The remainder of this entry focuses on Grenache Rouge, the most popular form of this grape. For a variety so prolific, it's known more as a blending grape than one that shines on its own, with some notable exceptions like at Château Rayas, the famed iconoclast estate in Châteauneuf-du-Pape, France, or Comando G, near Madrid in Spain. Grenache (or Garnacha) ripens late, thrives in warm climates, and tends to produce fruit-forward wines with high alcohol levels and low to moderate tannins. It's the basis of most

Côtes du Rhônes and an important blending grape in Rioja. For a time, the grape had a large presence in Australia, too. While that presence has dwindled, there are still some extraordinary one-hundred-plus-year-old ungrafted vines growing in Barossa worth tracking down. Cirillo Estate's "1850 Ancestor Vine" bottling is a benchmark, and Ochota Barrels' "The Green Room," made from eighty-year-old-vines, is terrific, too. In Oregon, winemaker Maggie Harrison makes a wine I love at Antica Terra. Thanks to California's Rhone Rangers (a group of wine growers committed to growing Rhône Valley varietals like Grenache, Syrah, and Mourvèdre), Grenache—along with other Rhône Valley varietals—gained acreage and market share in the late twentieth century. There are some beautiful varietal wines available today.

In blind tastings, Grenache (like Nebbiolo and Sangiovese) oxidizes easily, which means the red colors begin to turn slightly orange almost as soon as the wines are bottled. Grenache tends to be quite high in alcohol because the grapevines can handle heat, converting all of that sunlight into sugars. Grenache at its best reminds me of ripe strawberries, Provençal herbs, and sweet spices. Pair with everything from barbecue ribs to Peking duck, brisket, or even a ham sandwich. A hearty portobello sandwich would be delightful, too.

GRENACHE / GARNACHA (ROUGE) SNAPSHOT

COLOR: dark pink, with a garnet rim

ACIDITY: medium TA

ALCOHOL: 13.5 to 15 percent ABV

BODY AND MOUTHFEEL: full

TANNIN: low to medium—these wines can be really gentle and juicy!

FLAVORS AND AROMATICS: juicy strawberry, black plum, leather, tobacco, herbes de Provence, lavender

MALBEC

This grape, while originally from southwestern France, owes its popularity to Argentina, where it is the most-planted grape variety in the country. Malbec thrives in the high-altitude slopes of the Andes Mountains, where it benefits from sunny days and strong diurnal shifts (warm days and cool nights). Today, Argentina's vineyards account for 75 percent of its plantings by acre worldwide, even though historically Malbec grew throughout France in Bordeaux, the Loire Valley, and south, in Cahors. In Bordeaux, it served as an important grape until phylloxera ravaged the region. Malbec never took to grafting the way Merlot did, and today it's essentially a last-fiddle blending grape alongside Petit Verdot, intended to add color and structure to the more famous varietals Cabernet Sauvignon, Merlot, and Cabernet Franc. In Cahors, famous for its sunny days and limestone slopes, it is known as Cot and makes an inky wine.

By contrast, in Argentina, Malbec produces fruit-forward, balanced wines that are relatively low in tannins and acidity but full of bright, jubilant fruit—blackberries, cherries, pomegranates—plus milk chocolate and cocoa, and, in some of the more interesting examples, green tobacco, vanilla, and dill. Because Malbec has so few tannins, it can be a good match with leaner meats—roasted duck breast, lamb kebobs, and even barbecue chicken.

MERLOT

If you are someone who loves Merlot already, congratulations. And if you are someone who watched *Sideways* and decided it is not worthy of your attention, I invite you to reconsider that here. Merlot, as a grape, is like a big hug. It's soft and warm and juicy and delicious—and, fun fact, it is also a half-sibling to Cabernet Sauvignon, the most widely planted grape in the world. While Cabernet Sauvignon gets a lot of fanfare, Merlot is, in fact, the most-planted grape in France, where its vines outnumber Cabernet Sauvignon more than two to one. Merlot is best known for its soft tannins and deep plummy, slightly herbaceous flavor. In moderate climates like Bordeaux, it's also a smart bet for winemakers. Merlot ripens one to two weeks earlier than Cabernet Sauvignon, avoiding rain and problematic weather at harvest. As a result of its early ripening, Merlot is a grape of choice in most of the world's cooler to moderate climates, like northern Italy, Switzerland, Moldova, Columbia Valley (Washington), Chile, and, to a lesser extent, Argentina (where Malbec reigns supreme).

Merlot thrives in damp, clay-based soil. Pétrus, in Pomerol, on Bordeaux's Right Bank, makes stunning wines and the world's most famous Merlot-based example—and Masseto, from Tenuta dell'Ornellaia in Bolgheri, Tuscany, is a terrific Italian benchmark. But there are plenty of great, more accessible examples. Château Bourgneuf, in Pomerol, is a longtime sommelier favorite and makes gorgeous wines that retail for about sixty-five dollars per bottle. And, in the United States, former Le Bernardin sommelier Marie Vayron Ponsonnet (who grew up in Pomerol making wine at her family's estate, Château Bourgneuf) and Rajat Parr have teamed up in Napa Valley to make Colète, a gorgeous wine.

Pairing-wise, Merlot is a crowd-pleaser match with everything from a falling-off-the-bone roast lamb shoulder to a grilled rib eye. If you're opting out of meat, it's a delightful partner to sweet potatoes, squash, or sautéed kale—just expect some of the grape's greener, more herbaceous notes to shine through.

NEBBIOLO

Nebbiolo, a noble grape native to Piedmont, Italy, is sometimes called an "intellectual" variety because its charm is not always obvious until you've had it a few times. The grape, like the region it comes from, is dramatic and stunning—with naturally high acidity and tannins, an intensely perfumed nose, and a paradoxical combination of power and finesse. Aromas are often associated with rose petals, truffles, licorice, mushrooms, and dried cherries.

Nebbiolo is the grape of anything labeled Barolo or Barbaresco, the two most famous appellations in Piedmont. Structurally, Barolo and Barbaresco benefit from several years of aging, in barrel and bottle. Nebbiolo's name comes from *nebbia*, the Italian word for "fog," and is believed to allude to either the Piedmontese mist that covers the hillsides most mornings or a naturally occurring waxy coating on the grapes that resembles fog.

Great examples of Nebbiolo also exist in lesser-known Italian regions like Valtellina (where it is called Chiavennasca). Some people who love Nebbiolo are growing it outside of Italy, but, to date, the examples from elsewhere I've tried are not really my thing and almost seem like a shadow of Nebbiolo, without its soul.

Classic pairings with Nebbiolo include Piedmontese dishes like agnolotti del plin (tiny handmade stuffed pastas filled with veal, pork, or beef, and escarole) with shaved white truffle, carne cruda (finely chopped steak tartare), vitello tonnato (thinly sliced veal with a sauce of homemade mayonnaise and tuna), gnocchi al Castelmagno (gnocchi with a sauce of Castelmagno, a local semi-hard cow's-milk cheese), and brasato al Barolo (a dish of beef braised in a Barolo sauce). That said, don't be shy about opening a bottle paired with more accessible meals—Barolo with a good hamburger is fabulous, too.

NEBBIOLO SNAPSHOT

COLOR: garnet core with ruby flecks and a garnet to orange rim (depending on age)

ACIDITY: high TA

ALCOHOL: 13.5 to 15 percent ABV

BODY AND MOUTHFEEL: medium+

TANNIN: high

FLAVORS AND AROMATICS: dried cherry, red plum, violet, rose petals, tobacco, anise, tea leaf, crushed rock, black truffle, mushroom, licorice

PINOT NOIR

The term *Pinot Noir* comes from the French words for "pine" and "black" and alludes to the variety's tight, pinecone-shaped bunches of dark fruit. A thin-skinned grape, Pinot Noir generally produces wines that taste like tart cherries and happiness and is capable of transmitting terroir in a way that most other grape varieties are not.

Pinot Noir's homeland and global benchmark is Burgundy, where Romans cultivated vines as early as 1 CE and where Cistercian and Benedictine monks improved viticulture throughout the Middle Ages with resources from the wealthy and powerful Valois Dukes (see page 143), who bequeathed large swaths of land to the Church and invested significantly in the region's infrastructure.

Unlike most varieties, which are crossings or mutations of more ancient grapes, Pinot Noir is almost certainly a very old variety just a generation or two removed from wild vines. More than fifty Pinot Noir clones are legally recognized within France, twice the number of the much-more-planted Cabernet Sauvignon. Pinot Noir, head of a family of grapes known as Noiren, is first mentioned in print during the mid-fourteenth century, though it likely dates back much longer. The Noiren family is prone to mutation, evidenced by the array of grapes—Pinot(s) Blanc, Gris, and Meunier— that sprang from Pinot Noir. Today, compelling examples are produced throughout the globe, including in the United States, South Africa, New Zealand, and nearly everywhere else that cultivates vines. One of the reasons Pinot Noir can be so hard to get right is because it mutates so often and because it can be so challenging for vignerons to match a compatible clone with a compatible rootstock in a compatible plot of land. But when vignerons get it right, and the terroir is compelling, the wines are ethereal.

PINOT NOIR SNAPSHOT

COLOR: ruby and shades of dark pink

ACIDITY: medium to medium+ TA

ALCOHOL: 12 to 14.5 percent ABV

BODY AND MOUTHFEEL: medium+

TANNIN: medium+

FLAVORS AND AROMATICS: tart cherry, limestone, green tobacco, mushroom, violets, beetroot, tomato, anise, vanilla, baking spices, happiness

SANGIOVESE

Translated to "Blood of Jove" or "Blood of Jupiter" and named after the most important Roman god, Sangiovese is Tuscany's noblest grape, with the potential for creating incredibly long-aging and complex wines. Aromatic and structured, the grape thrives on limestone and a crumbly shale-clay soil known as galestro. In Montalcino, the "Brunello" clone is king. It grows south of Chianti, where the climate is warmer and the grape skin grows thicker, acting as sunscreen and bringing color and tannins along with it. As a result, Brunello produces fuller-bodied wines that benefit from longer aging and can continue to improve for decades.

The grape of Chianti, Brunello, Vino Nobile di Montepulciano, and many a Super Tuscan (wines made in Tuscany with disregard to appellation laws), Sangiovese is the most-planted grape in all of Italy. Its homeland is Tuscany, where various mutations form the basis for several famous regional wines. Sangiovese Grosso, for example, is a large and revered clone from which Brunello di Montalcino is made. Prugnolo Gentile is the clone associated with Vino Nobile di Montepulciano. Nielluccio is Corsica's name for this glorious grape. Sangiovese, like Nebbiolo, lacks swagger outside of Italy, though occasionally experimental winemakers plant it. Regardless of its incarnation, Italian Sangiovese always displays fresh tart-cherry characteristics, sun-dried tomatoes, savory herbs (thyme and oregano, for me), and a bright, piercing acidity.

Brunello di Montalcino and Rosso di Montalcino are both Sangiovese Grosso–based wines from the Montalcino region, in southern Tuscany. Brunello is required to spend more time aging before it is released, so the tannins have had time to soften. Rosso di Montalcino is like a fruitier "baby Brunello" with less aging and a lower price tag.

Sangiovese is an epic pairing with pasta all'Amatriciana (where it matches the brightness of the tomato and is a perfect counterpoint to the fatty pancetta), a grilled bistecca alla Fiorentina, or a plate of cured meats like prosciutto and bresaola and some Parmesan cheese.

SANGIOVESE SNAPSHOT

COLOR: red, with orange and purple flecks and a garnet rim (depending on the amount of time aged in oak)

ACIDITY: high TA

ALCOHOL: 13 to 14.5 percent ABV

BODY AND MOUTHFEEL: medium+

TANNIN: high

FLAVORS AND AROMATICS: ripe and dried tomato, tart cherry, rosemary, thyme, oregano, leather, baked earth

SYRAH

A noble grape born from the obscure grape varieties Dureza and Mondeuse Blanche, Syrah originated in southern France, forged its identity in the Rhône Valley, and now flourishes in vineyards throughout the world, from Australia to Uruguay. It is known for producing long-lived, nuanced red wines with a particular savory aroma that is beguiling and majestic in the best examples. A fun fact: Syrah is related to Pinot Noir (it is Pinot Noir's "great-grandchild").

In the eighteenth and nineteenth centuries, Syrah from the tiny appellation of Hermitage was so prized that wine merchants shipped it up the Rhône River to Bordeaux, then blended it with the paler red wines of the Médoc to add color and structure in a process known as Hermitagé—a practice eventually outlawed in the nineteenth century. Benchmark Syrahs from the Northern Rhône can be inky and purple in their youth, medium to full in body, and redolent of red and black fruit, violets, leather, black pepper, and grilled meat. In Australia, South Africa, and now elsewhere, Syrah is called by its Australian name, Shiraz. Some of the oldest, ungrafted vines in the world grow in Australia's Barossa Valley, thanks to a man named John Macarthur, who is believed to have transported vines there from Europe in 1817.

Syrah's popularity has skyrocketed in recent decades. Areas like Red Mountain in Washington State show momentum and promise, while an initial popularity surge in California at the turn of the twenty-first century has hit the brakes. While the grape needs a warm climate to fully ripen, temperatures that are too hot can interfere with Syrah's distinctive aromas, and its identity can get lost. Syrah is especially popular and prominent in South

Africa, where it is also generally called Shiraz. While much of New Zealand is too cool for the grape to ripen, Hawke's Bay is having success and showing exciting potential.

Syrah is a classic match with roasted lamb and game birds and meat like squab, venison, or partridge. It's also a great idea with something as simple as a roast chicken with rosemary potatoes and a side of broccoli rabe. Northern Rhône Syrah always reminds me of a good pastrami sandwich, so I would be remiss not to suggest that pairing, too.

TEMPRANILLO

The main grape of Rioja, Tempranillo is also Spain's noblest and most-planted red grape—and the fourth-most-planted worldwide. Tempranillo is grown throughout the globe, though primarily in Iberia, where it has been cultivated since the time of Phoenician settlements as far back as 1500 BCE. Tempranillo is so prominent that it has dozens of regional names throughout Spain, including Ull de Llebre, Cencibel, Tinto Fino, and both Aragonez and Tinta Roriz (in Portugal).

Tempranillo's name comes from the Spanish word *temprano* ("early")—as the grape ripens approximately two weeks earlier than Garnacha, the second-most-planted grape in Spain and a compatible counterpart to Tempranillo in many wines, including Rioja. Composition-wise, it is a thick-skinned black grape variety with a relatively neutral profile. Tempranillo is low in both acidity and sugar, meaning it usually doesn't get overly tart nor high in alcohol, even in hot, arid countries like Spain. Wines made from Tempranillo tend to be juicy with soft, supple tannins and ripe red fruit. They're usually aged in new oak, often American (thanks to Spain's early ties with the Americas and its historical beef with France), imparting sweet vanilla and herbaceous dill flavors.

ZINFANDEL

While America gave this grape its star power, Zinfandel's roots are European. Known as Primitivo in Puglia (the "heel" of Italy's boot), modern DNA testing eventually confirmed that the variety originated in Croatia, where it's known as both Tribidrag (*trib-ee-drag*) and Crljenak Kaštelanski (*ser-ee-AY-lack cash-tel-on-skee*), thanks to previous confusion that it was two separate varieties.

In America, Zinfandel thrives in warm regions and is best associated with California, where it has a long history dating back to the 1820s, when it is believed to have arrived in someone's suitcase from the Austrian Imperial Nursery. Zinfandel's prolific tendencies helped the grape soar in popularity by the late nineteenth century, when quantity was prized above all else.

Ironically, today the best examples are from old, ungrafted vines planted more than a century ago. Lodi, south of Sacramento and southeast of Napa Valley, is home to some of the world's most striking vineyards—more than one-hundred-year-old gnarly, beautiful, ungrafted Zinfandel vines growing in sand. (If you visit, you'll learn that for years, and even today, Zinfandel has been the unspoken "secret ingredient" in many a Napa Valley Cabernet Sauvignon, which must include just 75 percent of a stated *Vitis vinifera* grape in order to be labeled as such.

Award-winning author, photographer, and former sommelier Randy Caparoso is perhaps Zinfandel's greatest advocate and educator when it comes to these old vines. (He literally wrote the book on the region, called *Lodi! The Definitive Guide and History of America's Largest Winegrowing Region.*) Paul Draper and Ridge Vineyards (which first vinified Zinfandel in 1966) from the Geyserville vineyard in Sonoma, California, proved that extraordinary wines could come from the grape. Today, Geyserville vines are over 130 years old. Tegan Passalacqua, proprietor of Sandlands Vineyards and full-time winemaker at Turley Wine Cellars, and Morgan Twain-Peterson of Bedrock Wine Co. are the next generation to advance Zinfandel's identity through their beautifully crafted wines.

Zinfandel wines can be high in alcohol, thanks to the grape's elevated sugar content, but well-managed examples are balanced and beautiful. Zinfandel is also known in "white" form, stemming from an accident in 1972 in which a tank of saignée Zinfandel at the Sutter Home winery got stuck and wouldn't ferment to dryness. The winery dosed it with sulfur and bottled it anyway—it was pink and sweet, people loved it, and the rest is history.

My favorite Zinfandel pairings are anything with a touch of sweetness or spice—like brisket, Thai food, roasted pork shoulder, or honey-glazed ham. Maple pecans would probably be great, too, if you're in a snacking mood.

ZINFANDEL SNAPSHOT

COLOR: ruby, with a pink rim

ACIDITY: medium to medium+ TA

ALCOHOL: 14 to 17 percent ABV

BODY AND MOUTHFEEL: full

TANNIN: medium+

FLAVORS AND AROMATICS: fresh ripe strawberry, peach yogurt, ripe cherry, dried fig

CHEAT SHEET TO NAVIGATING A WINE LIST

As we'll cover in the next chapters, the "Old World" is one in which grapes and regions are inextricably linked. Labeling wines by grape variety is an American invention and isn't yet one hundred years old. (Ha, plenty of Lodi Zinfandel is older!) The tradition of labeling wines by grape varietal began in California in 1935, just after Prohibition, and resulted from California growers beginning to understand which grapes had the potential to make excellent wines and which did not, along with some help and nudging from the quality-minded first head of UC Davis's department of viticulture and enology, Professor Maynard A. Amerine, after the department reopened post-Prohibition.

To make navigating those "Old World" regions a bit easier, below is a cheat sheet so you can remember which grapes correspond to which appellations. It's not all-inclusive, but hopefully it's a helpful starting place. If you are keen for a comprehensive list, the answers are easy to google—or you can just befriend anyone studying for wine exams. This overview is a natural segue as we head into the next two chapters, which focus on major wine-producing countries in both the "Old" and "New Worlds."

SPECIFIC REGIONAL NAME	COUNTRY & GENERAL REGION	GRAPES
ALSACE (WHITE)	France	Labeled by grape, unless it's a field blend (marked "Edelzwicker"). The four official noble grapes of Alsace are Muscat, Riesling, Pinot Gris, and Gewürztraminer, which can be used to make Grand Cru bottlings indicated as "Alsace Grand Cru." The Sylvaner grape is allowed as an exception in the Grand Cru vineyard Zotzenberg.
AMARONE (RED)	Italy (Lombardy)	Wine made from ripe, then raisinated Corvina, Rondinella, and Molinara grapes
BEAUJOLAIS (WHITE)	France	100% Chardonnay
BEAUJOLAIS (RED)	France	At least 90% Gamay (up to 10% other grapes, including Chardonnay, Pinot Gris, and Aligoté)
BORDEAUX (WHITE)	France	Sauvignon, Sémillon, and Muscadelle
BORDEAUX (RED)	France	Cabernet Sauvignon, Merlot, Cabernet Franc, Petit Verdot, Carmenère, and Malbec are permitted , with up to 5% combined others
BORDEAUX (ROSÉ)	France	A blend of mainly Cabernet Sauvignon, Cabernet Franc, Malbec, Merlot, Petit Verdot, and Carmenère
BOURGUEIL (RED)	France (Loire Valley)	Cabernet Franc, with up to 10% Cabernet Sauvignon

BRUNELLO DI MONTALCINO (RED)	*Italy (Tuscany)*	Sangiovese (called Sangiovese Grosso here)
BURGUNDY (RED)	*France (Chambolle-Musigny, Gevrey-Chambertin, Morey-Saint-Denis, Volnay, and Vosne-Romanée)*	Pinot Noir
BURGUNDY (WHITE)	*France (Chassagne-Montrachet, Puligny-Montrachet, and Meursault)*	Chardonnay, Pinot Blanc, and up to 30% Pinot Gris
CHABLIS (WHITE)	*France (Burgundy)*	Chardonnay
CHAMPAGNE (WHITE AND ROSÉ—AND ALWAYS SPARKLING)	*France*	Chardonnay, Pinot Noir, and Pinot Meunier are the three main grapes
CHÂTEAUNEUF-DU-PAPE (WHITE)	*France (Southern Rhône)*	Grenache Blanc and Gris, Bourboulenc, Roussanne, Clairette, Clairette Rose, Muscardin, Picardan, Piquepoul Blanc and Gris, and others
CHÂTEAUNEUF-DU-PAPE (RED)	*France (Southern Rhône)*	Grenache Noir, Mourvèdre, Syrah, Cinsault, Counoise, Terret Noir, Piquepoul Noir, and others
CHIANTI (RED)	*Italy (Tuscany)*	Sangiovese, Canaiolo, Ciliegiolo, and others
CHINON (RED)	*France (Loire Valley)*	Cabernet Franc, with up to 10% Cabernet Sauvignon
CORNAS (RED)	*France (Northern Rhône)*	Syrah
CÔTES DU RHÔNE (RED)	*France (Northern and Southern Rhône)*	Grenache, Mourvèdre, Carignan, and others
GRAVES (WHITE)	*France (Bordeaux's Left Bank)*	A blend of mostly Sémillon and Sauvignon Blanc, with Sauvignon Gris and Muscadelle also allowed
GRAVES (RED)	*France (Bordeaux's Left Bank)*	Cabernet Sauvignon, Cabernet Franc, Merlot, Petit Verdot, Malbec, and Carmenère
GRAVES SUPÉRIEURES (OFF-DRY OR SWEET WHITES)	*France (Bordeaux's Left Bank)*	Botrytis-infected or late-harvest raisinated Sémillon, Sauvignon Blanc, Sauvignon Gris, and Muscadelle

HERMITAGE BLANC (WHITE)	*France (Northern Rhône)*	Marsanne and Roussanne blend
HERMITAGE ROUGE (RED)	*France (Northern Rhône)*	Syrah
MÂCON (WHITE)	*France (Burgundy)*	Chardonnay
MARGAUX (RED ONLY)	*France (Bordeaux's Left Bank / Médoc)*	Cabernet Sauvignon, Cabernet Franc, and Merlot-dominant blend with Petit Verdot, Malbec, and Carmenère
MÉDOC (RED ONLY)	*France (Bordeaux's Left Bank)*	Cabernet Sauvignon, Cabernet Franc, and Merlot-dominant blend with Petit Verdot, Malbec, and Carmenère
MENETOU-SALON (WHITE)	*France (Loire Valley)*	Sauvignon Blanc
MENETOU-SALON (ROSÉ, RED)	*France (Loire Valley)*	Pinot Noir
MEURSAULT (WHITE)	*France (Burgundy)*	Chardonnay
PAUILLAC (RED ONLY)	*France (Bordeaux's Left Bank / Médoc)*	Cabernet Sauvignon–dominant blend, with Merlot, Cabernet Franc, and Petit Verdot
POMEROL (RED ONLY)	*France (Bordeaux's Right Bank)*	Merlot-dominant blend with Cabernet Franc as the main supporting grape; Cabernet Sauvignon, Malbec, and Petit Verdot also allowed in blends, in small quantities
POUILLY-FUISSÉ (WHITE)	*France (Burgundy)*	Chardonnay
POUILLY-FUMÉ (WHITE)	*France (Loire Valley)*	Sauvignon Blanc
QUINCY (WHITE)	*France (Loire Valley)*	Sauvignon Blanc
RÍAS BAIXAS (WHITE)	*Spain (Galicia)*	At least 70% Albariño (Treixadura and Loureiro also allowed); 100% when labeled as "Albariño"
RIBERA DEL DUERO (RED)	*Spain (Castilla y León)*	At least 75% Tempranillo (Garnacha, Malbec, Merlot, and Cabernet Sauvignon also allowed)
RIBERA DEL DUERO (WHITE)	*Spain (Castilla y León)*	At least 75% Albillo Mayor (the only permitted white grape in the region)

RIOJA (WHITE)	*Spain (La Rioja)*	Viura, Chardonnay, Sauvignon Blanc, Verdejo, Malvasia, Garnacha Blanca, and others
RIOJA (RED)	*Spain (La Rioja)*	Tempranillo, with Grenache, Mazuelo, Graciano, and Maturana
SAINT-ÉMILION & SAINT-ÉMILION GRAND CRU (RED)	*France (Bordeaux's Right Bank)*	Merlot and Cabernet Franc–dominant blend, with Cabernet Sauvignon, Malbec, Carmenère, and no more than 10% Petit Verdot also allowed
SAINT-ESTÈPHE (RED ONLY)	*France (Bordeaux's Left Bank / Médoc)*	Blend of Cabernet Sauvignon, Cabernet Franc, Merlot, Petit Verdot, Malbec, and Carmenère, although generally mostly Cabernet Sauvignon, with Merlot and Cabernet Franc as main supporting players
SAINT-JOSEPH (RED)	*France (Northern Rhône)*	Syrah, with up to 10% Marsanne and Roussanne
SAINT-JOSEPH (WHITE)	*France (Northern Rhône)*	Marsanne and Roussanne
SAINT-JULIEN (RED ONLY)	*France (Bordeaux's Left Bank / Médoc)*	Blend of Cabernet Sauvignon, Cabernet Franc, Merlot, Petit Verdot, Malbec, and Carmenère, although generally mostly Cabernet Sauvignon, with Merlot and Cabernet Franc as main supporting players
SAINT-VÉRAN (WHITE)	*France (Burgundy)*	Chardonnay
SANCERRE (WHITE)	*France (Loire Valley)*	Sauvignon Blanc
SANCERRE (RED AND ROSÉ)	*France (Loire Valley)*	Pinot Noir
SAUMUR, SAUMUR-CHAMPIGNY (RED)	*France (Loire Valley)*	Cabernet Franc and up to 30% combined Cabernet Sauvignon and Pineau d'Aunis
SAUTERNES (SWEET)	*France (Bordeaux's Left Bank)*	Hand-harvested, botrytis-infected grapes, mainly Sémillon and Sauvignon Blanc, with Sauvignon Gris and Muscadelle allowed
SHERRY (FORTIFIED—DRY AND SWEET)	*Spain (Xeres)*	Palomino and others
SOAVE (WHITE)	*Italy (Piedmont)*	Garganega, Trebbiano di Soave, and others
VALPOLICELLA (RED)	*Italy (Lombardy)*	Corvina, Molinara, Rondinella, and others

CHAPTER 6

"OLD WORLD" COUNTRIES AND REGIONS

While the lines between the "Old World" and "New World" are becoming increasingly blurred, thanks to globalization, the "Old World" still has a several-thousand-year head start in viticulture. Remember that *Vitis vinifera* grapes could not even grow in the United States were it not for a series of blunders and subsequent scientific discoveries made little more than one hundred years ago (see A Brief History Lesson on page 21).

This chapter is intended as a high-level overview, a greatest hits of countries and subsequent regions you're likely to encounter, with a nod to the grapes usually grown there. Countries are listed alphabetically in this chapter and the next, but regions are organized by their north-to-south geographic location. This was helpful to me when I began my journey trying to understand how grapes and regions all fit together. This chapter has some overlap with chapter 5, though it's presented through the lens of regions and countries, while the previous chapter focuses on getting to know the most prominent grapes.

CONNECTING PLACES AND GRAPES

I did not know that Sancerre was a region in France, let alone a subregion of the Loire Valley's Central Vineyards, when I sold my first glass of it to a guest early in my wine career. Sancerre had transcended any geographic location, an entity unto itself, sort of like Madonna—without a binding prior identity. Wines from France, Italy, and generally most of the Mediterranean Basin have long been labeled by region; each region has a long history with the grapes that grow best there and they have been codified into wine law. Wines from the United States and other "New World" countries are labeled primarily by grape variety, because these countries are still getting the hang of what grapes grow best where (and possibly, in the United States, because we have never been great at following other people's rules). Remember, labeling wines by grape variety is an American invention dating back only to the mid-1930s!

AUSTRIA

Records of Austria's wine history date back to the fifteenth century BCE, when the Romans controlled the land. But viticulture was likely introduced (without record) a few centuries before that, by Celtic tribes. Austria, like anywhere with a long history in wine, has experienced its share of speed bumps, ranging from an edict in 92 CE from the Roman emperor Domitian, who ordered all vineyards pulled up in favor of grains (he wanted to be sure his armies were fed), to various taxes and regulations and two world wars. More recently, in a 1985 scandal, a bad-apple chemist encouraged a handful of growers to add diethylene glycol (DEG) to a few dilute vintages to make them appear sweeter and fuller-bodied (toxic to humans if consumed, DEG is water-soluble and commonly added to beauty products and antifreeze). Luckily the government realized what he was doing and clapped back with some of Europe's strictest quality-oriented wine laws.

A history of excellence is deeply ingrained in Austria, too, beginning in the eighth century BCE with Charlemagne's decision to impose rules for wine production, which he learned from Cistercian monks in nearby Burgundy. And in 2001, in reaction to the DEG scandal, Austria implemented the Districtus Austriae Controllatus system (DAC—like France's AOC/AOP), which regulates quality wine using extremely strict standards by geographic location. Austria's standards today are among the highest anywhere—it produces just 1 percent of the world's wine, and over 90 percent of that is classified as Qualitätswein (quality wine).

Austria's vineyards yield primarily white grapes, with Grüner Veltliner accounting for nearly 33 percent of vineyard land. While this grape's existence was first recorded in 1855 (just a blink on the timeline of winemaking), it wasn't particularly popular until the 1960s and 1970s, thanks to its ability to generate quantity during a time when that was especially prized. (Fortunately for everyone, quality didn't suffer!) Riesling is also an important grape and is a star player, especially on some of the Wachau's steep slopes. Chardonnay (called Morillon here) and Sauvignon Blanc are quite popular, too, in the southerly region of Styria, or Steiermark.

NIEDERÖSTERREICH

The Niederösterreich, or Lower Austria, encompasses the country's northeast corner and is responsible for more than half of its wine production. There are eight subregions in the area, and all of them are DACs. The three most noteworthy are the Wachau, the Kremstal, and the Kamptal regions, all of which specialize in Grüner Veltliner and Riesling. While much of Austria is considered a continental climate, Lower Austria's DACs border the Danube River, which has varying degrees of impact on the individual regions but, in general, helps moderate temperatures in the area and reflects light up onto the vines of the region's steepest vineyard slopes.

AUSTRIA
WINE REGIONS

CZECHIA

GERMANY

Kamptal
Kremstal
Krems
Wachau

Niederösterreich

Danube

● **Salzburg**

Vienna

Neusiedlersee

Burgenland

HUNGARY

Steiermark

SLOVENIA

ITALY

CROATIA

MEDITERRANEAN SEA

THE KAMPTAL AND THE KREMSTAL

The northern regions of the Kamptal and the Kremstal are also known for exceptional Grüner Veltliner and Riesling, with riper, often botrytized styles coming from the Kremstal (which is warmer, more humid, and less windy) and racier ones from the Kamptal (which is drier and benefits from cool nightly winds from the forests to its west). There are so many terrific producers here, too. Some of my favorites include a bit of crossover with the Wachau, which makes things easier. Top on my list are the Wachau-based Emmerich Knoll, F.X. Pichler, Veyder-Malberg, Hirsch Vineyards, Bründlmayer, and Sohm & Kracher.

THE WACHAU

Even the name, the Wachau (*VOCK-ow*), sounds commanding—and it should; these are wines worth your attention! In the Wachau, the region with the longest history of excellent winemaking in Austria, vines are grown on steep, terraced slopes with a soil of granite and decomposing calcareous loess, swept in from the Dolomites over millions of years. Hand-built granite walls keep the vineyards on the hillsides, and the region is now a UNESCO World Heritage Site, with viticultural history dating back to 830 CE. Austria's most powerful and famous expressions of both Grüner Veltliner and Riesling are grown here, where most vineyards are planted on the northern bank of the Danube and benefit from long sunny days and south-facing slopes. Despite the sunshine, northerly winds rush down the river valley, cooling the vines and providing excellent brightness and acidity. Basically, the Wachau is one of nature's wine miracles.

In the Wachau, wines are classified according to ripeness levels: Steinfeder, Federspiel, and Smaragd. Steinfeder, the lightest in style, may not reach more than

SEKT (SPARKLING WINE)

Sekt (the term used in Germany and Austria for sparkling wine) has a history dating back to the mid-nineteenth century. Since 2016, Austria has gotten serious about Sekt, passing a law creating four separate categories. Three of the four categories are recognized as a PDO, or Protected Designation of Origin, the EU's categorization for quality wine. The top two quality tiers (Sekt Austria Grosse Reserve PDO and Sekt Austria Reserve PDO) must be made from hand-harvested, whole-cluster–pressed grapes made in the traditional method (see page 51), then aged on the lees for thirty-six and eighteen months, respectively. The third tier, Sekt Austria PDO, is more lenient, allowing any production method and requiring just nine months of lees aging. The fourth category is non-PDO Sekt, so it doesn't have to meet any of these quality standards. I will not be rushing to explore those myself, but I'm sure my younger self would have been glad to give them a try!

Germany is well known for Sekt, too. If you want a quality guarantee, look for bottles labeled "Winzersekt," which come from estate-grown single-vineyard grapes and are made in the traditional method.

11 percent alcohol by volume. Federspiel wines have between 11.5 and 12.5 percent ABV. Smaragd wines are the fullest in body, with at least 12.5 percent ABV. On Austrian labels, *Reid* means "vineyard site"—the region has a long history of honoring vineyard sites, thanks to its exceptional terroir and monastic roots. Some favorite producers include Emmerich Knoll, F.X. Pichler, Rudi Pichler, Prager, and Veyder-Malberg.

THE BURGENLAND

South of Lower Austria is the Burgenland, bordering Hungary. Records suggest that winemaking here dates back to the sixteenth century, but the wine industry didn't hit its stride until the 1950s, once the area joined the First Austrian Republic after World War II and Russian occupation ended. The region is Austria's warmest and benefits from the large, shallow Lake Neusiedl (*noy-SEED-el*). The lake creates a perfect, humid environment for botrytis. One Burgenland region, Neusiedlersee, is famous both for its dry red wines made from the Blaufränkisch grape as well as botrytized sweet wines made from a variety of both red and white grapes, especially the white grape Welschriesling (which, despite its name, has no relation to Riesling).

FRANCE

Viticulture in France dates back to the fifth century BCE when Greek settlers arrived and brought winemaking techniques with them. For the first thousand years of its existence, France's wine scene relied on foreign trade. Once the Gauls (people who inhabited France around 400 BCE) got into their groove, they started growing and pruning their own vines and then, meaningfully, did something no one had done beforehand—they began to treat winemaking as an art. Their wines became very good. Before long, they were world-famous.

The Middle Ages were a golden era for wine in France. In Burgundy, monasteries had the resources for and interest in making exceptional wines, and the wealthy and powerful Valois Dukes (our guy—Philip II the Bold!) prioritized it. France's wine scene continued to gain momentum through the 1855 World's Fair, held in Paris, and up until the late 1800s, when phylloxera and downy mildew ravaged Europe's vineyards, followed in short order by World Wars I and II.

France was also the first country to codify wine laws. They created the AOC system—now known as the AOP system, or Appellation d'Origine Protégée—in 1935. This system essentially determines the geographical boundaries of each region and regulates which grapes are allowed to grow within them. (It also limits

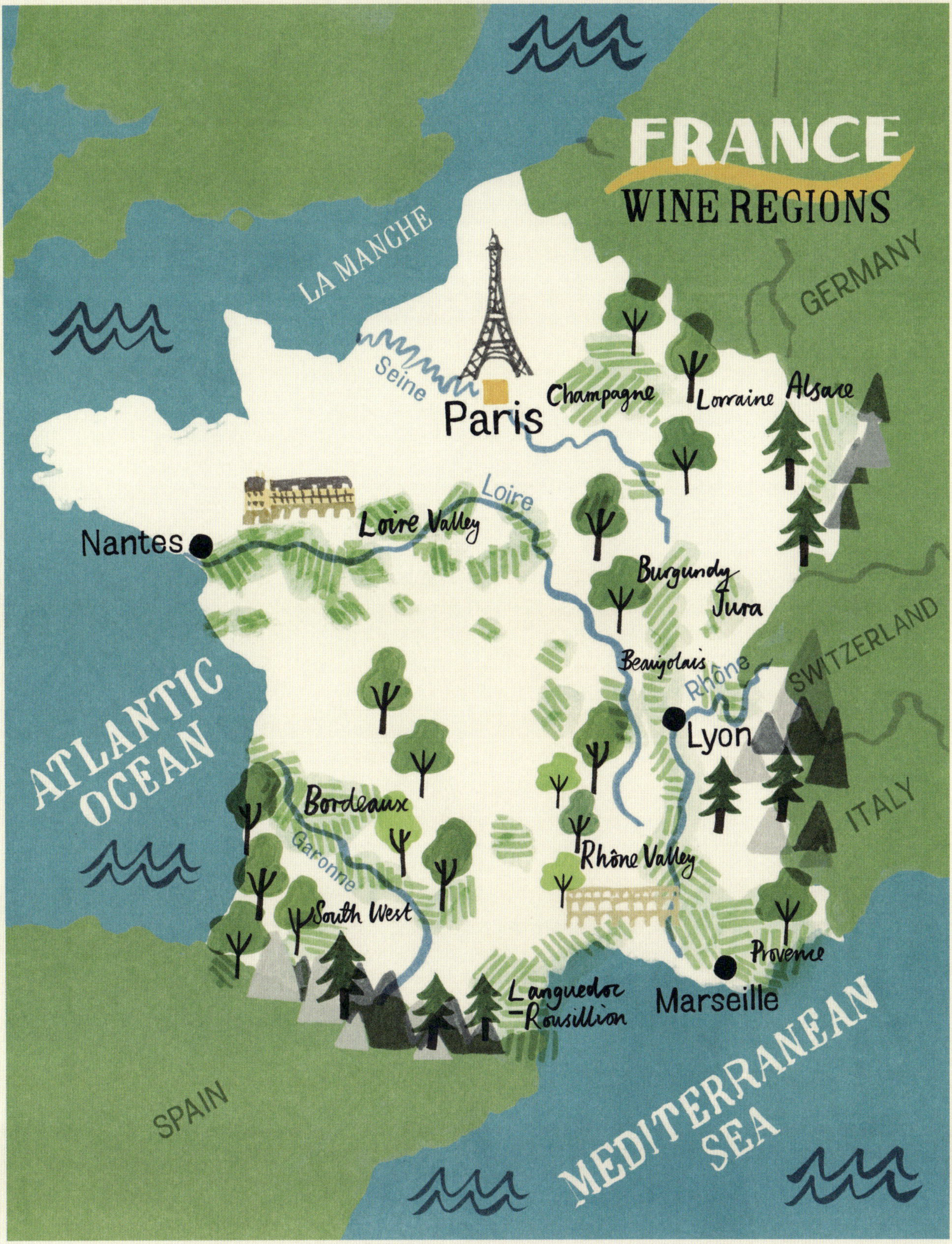

FRANCE
WINE REGIONS

ATLANTIC OCEAN

LA MANCHE

GERMANY

SWITZERLAND

ITALY

SPAIN

MEDITERRANEAN SEA

Seine

Paris

Champagne

Lorraine

Alsace

Nantes

Loire

Loire Valley

Burgundy

Jura

Beaujolais

Rhône

Lyon

Bordeaux

Garonne

South West

Rhône Valley

Languedoc-Rousillion

Provence

Marseille

grape yields and vineyard density, training and pruning techniques, production methods, minimum alcohol levels, and minimum must weight at harvest.) But the main takeaway here is that the AOP decided certain regions were best suited to certain grapes, and that the region—or named place—sits at the top of the hierarchy.

CHAMPAGNE

Champagne is divided into five subregions. The three most prominent are the Montagne de Reims, to the north, where Pinot Noir flourishes; Vallée de la Marne, to the west, where Pinot Meunier reigns supreme (this grape successfully grows on sites too cold for Pinot Noir and Chardonnay and also ripens earlier); and the Côte des Blancs, named for both its white, chalky slopes and for Chardonnay, the region's most famous white grape, which thrives there. Two other subregions, the Côte de Sézanne (south of the Côte des Blancs) and the Côte des Bar (farther south still), are less famous than their northerly neighbors but are producing more and more terrific examples. Climate change brings the benefits of warmer weather and ripe, juicy grapes in this historically cool climate, and a trend of ambitious young growers eager to vinify their family's vines, instead of selling them to larger houses, is benefiting everyone who loves delicious sparkling wines. A few of my favorite producers throughout Champagne are listed in the appendix (page 232). Sparkling

Among the Grande Marque ("big brand") Champagnes, which include large established houses, just one—Louis Roederer's Cristal—stands apart in grape quality and vineyard work. A blend of 60 percent Pinot Noir and 40 percent Chardonnay, Roederer's Cristal is the only Grande Marque made entirely of estate-owned, biodynamically farmed grapes since 2012. Cristal was initially commissioned in 1876 for Czar Alexander II of Russia, who requested a clear glass bottle with no punt (indentation at the bottom) because he was paranoid about being bombed or poisoned. The bottle has remained clear throughout its history, showing off the deep gold color of the wine, which is aged over six years and always has a vintage date. Unlike most Champagne houses, Roederer owns a majority of their own vineyards and has an intimate understanding of the best parcels. After the czars came to an end with the Russian Revolution in 1917, Roederer continued to produce Cristal. The house began releasing it commercially in 1945.

wines from Champagne must be made in the traditional method (see page 51) from grapes grown within its designated appellation. While Chardonnay, Pinot Noir, and Pinot Meunier are the three dominant grapes in the region, seven are allowed. The lesser-known four are Petit Meslier, Pinot Blanc, Pinot Gris, and Arbane—information that is unlikely to come in handy but is fun wine trivia nevertheless.

139

ALSACE

If Alsatian wines are not yet among your favorites, I am excited for you to discover them. This region has belonged to both France and Germany throughout history, and the regional grapes reflect that. Riesling, Gewürztraminer, Pinot Gris, and Muscat are the four officially recognized noble grapes grown here, often bottled in tall green- or gold-colored glass bottles reminiscent of German Riesling. Alsace is located just east of the Vosges Mountains and benefits climatically from a "rain shadow" effect: storms form and get trapped on the western side of the mountains, leaving blue skies and plenty of sunshine for Alsace's slopes. Soils are a tapestry of limestone, sandstone, granite, schist, clay, loess, chalk, and volcanic sediment. While many wines are single-varietal, others are field blends of different grapes co-planted in a vineyard. Bottles labeled with the term "Edelzwicker" are a blend. Alsace makes sparkling wine (Crémant d'Alsace) and still white, red, and rosé, as well as delicious sweet wines indicated by "Vendange Tardive" or the letters "VT." Alsace also has fifty-one officially recognized Grand Cru vineyards throughout the region. All wines labeled as Grand Cru must list the vineyard on the label and be made from exclusively the Alsatian "noble varietals," with the exception of Zotzenberg, which can also include Sylvaner, and Vorbourg, Kirchberg de Barr, and Hengst, which can be made from Pinot Noir.

Champagne's Humble Beginnings (and Britain's Role)

Champagne started off as a bit of an outcast region before a few happy accidents—a newly crowned king, a British scientist, and a party-boy duke who breathed popularity into this new sparkling style with eighteenth-century grassroots influencer marketing.

North of (and therefore cooler than) Burgundy, which had similar limestone soils and specialized in exactly the same grapes (Pinot Noir and Chardonnay), Champagne produced still, tart, pale, and sad versions of its more famous southerly neighbors from the fifth century, when Romans brought vines to the area, until the eighteenth century. (Ruinart, the oldest surviving Champagne house to make exclusively sparkling wine, was established in 1729.) While cooler temperatures are a detriment to making still wine, they are the very thing that accidentally spurred the sparkle. Cooler temperatures meant that grapes got a late start to ripening, and temperatures dropped earlier in the autumn. When temperatures drop below about 65 degrees Fahrenheit, yeasts get tired and go to sleep. When yeasts are dormant, fermentation stops. Louis Pasteur did not discover yeast's involvement until the nineteenth century, so wineries prior to that bottled their wines thinking fermentation had finished, only to have yeasts reactivate the following spring when the weather warmed.

We can thank British scientist Christopher Merret (1614–1695) for figuring out how to safely transform still wine into sparkling wine. Merret was the first person to catalog the process of adding sugar and molasses to wine to intentionally create a secondary fermentation, resulting in wine that was "brisk and sparkling," back in 1662—fifty-six years before the French produced its sparkling wine treatise in 1718.

Around 1630, fellow Brit Sir Kenelm Digby, who owned a glassworks company, began using coal instead of wood (to make the flames hotter) and added more sand into his glass than potash and lime. This created bottles so much stronger than and superior to previous bottles that the Brits soon gained a monopoly on glass that could withstand the pressure of sparkling wine. (Prior to this, sparkling wine exploded in cellars throughout Champagne, wreaking havoc and causing a number of untimely deaths.)

Party boy Philippe II, Duke of Orléans, can be credited with the grassroots marketing campaign that raised sparkling Champagne's popularity in France. He loved it and made it a staple at his extravagant Parisian parties in the early eighteenth century. Nobility, press, and influencers of the day were intrigued, and the rest is history.

LOIRE VALLEY

Located southwest of Paris, the Loire Valley is one of France's largest and most diverse regions, following the (extremely long) Loire River from Central France west to the Atlantic. Still red, white, and rosé wines are made here, as are excellent sweet and sparkling bottles. The region abounds with quality and diversity. And while it feels unfair to try to cram this region (or any other) into such a small section, it's the more reason for you to start here and then follow your curiosity as far as you'd like.

Key subregions, from east to west, are listed below.

CENTRAL VINEYARDS (OR, HOME OF SANCERRE AND ITS NEIGHBORS)

This region is named Central Vineyards because the land is so far east it's located in central France. The primary grapes here are Sauvignon Blanc (white) and Pinot Noir (red, rosé), and the most famous appellation is Sancerre. Its lesser-known neighbors include Menetou-Salon and Quincy—both produce Sauvignon Blancs in a similar style with perhaps less complexity but markedly lower price points.

Pro tip #1: Look for wines labeled "Les Monts Damnés" on Sancerre bottlings. This indicates the grapes were grown on the steep "Damned Mountains" in the village of Chavignol. The soil here is prized Kimmeridgian marl, and the slopes are so steep they are "damned."

Pro tip #2: Domaine Vacheron wines have been among my favorite in the world since I first tried them decades ago. Winemakers Jean-Dominique and Jean-Laurent Vacheron have long farmed all of their vines biodynamically and are as meticulous and thoughtful as anyone. If you are collecting special bottles, keep an eye out for the domaine's own-rooted bottling, L'Enclos des Remparts, a Sancerre made from ungrafted Sauvignon Blanc vines.

TOURAINE (HOME OF VOUVRAY, PLUS SOME LOVELY REDS!)

Touraine is west of the Central Vineyards and home to Vouvray (Chenin Blanc in all its glory, from laser-sharp and dry to sparkling to late-harvest and fabulously sweet). It's also home to the red grape Cabernet Franc, which you'll find in bottles labeled Chinon, Bourgueil, and Saint-Nicolas-de-Bourgueil. The soil here is tuffeau, a kind of porous, chalky limestone that results in notably mineral expressions of these grapes in the final wine.

ANJOU-SAUMUR (OR, CHENIN BLANC IN ALL ITS FORMS AND SPLENDOR)

West of Touraine, Anjou-Saumur offers the full gamut of Loire wines and is the center for its sparkling bottles. Chenin Blanc is the star here, with regions like Savennières producing extraordinary examples. Savennières sits on the north bank of the Loire, on soil of decomposing schist. The vines soak up sunshine reflected off of the river and the results are ripe, rich, extraordinary examples of Chenin Blanc. On the opposite bank, south of the Loire, lies the Coteaux du Layon, known for sweet wines made from late-harvest Chenin Blanc grapes harvested by hand. These berries can be botrytized (affected with noble rot), lending more acidity, richness, and those famous notes of ginger, saffron, and chamomile.

PAYS NANTAIS (OR, LAND OF MUSCADET)

Muscadet dominates this far-west region, something that's been true since 1709 when a frost wiped out all of the red grapes. Most of the wines on shelves are labeled "Muscadet Sèvre et Maine," a subregion named for the confluence of two rivers. Wines are made from the grape Melon B (see page 113) and often made sur lie—spending time on the lees before they are bottled. These wines are delightfully citrusy and slightly saline and are a classic pairing with oysters (but also great anytime you're in the mood for something crisp and unfussy).

142

BURGUNDY

Located in central-eastern France, about three hours southeast of Paris, Burgundy is the poster child of "continental climate" wine regions throughout the world. It is located on the gentle limestone slopes of the Saône Valley, a natural fault line that emerged when the Alps and Pyrenees pushed up into the skies. Since Burgundy is typically quite cool in winter and hot during the summer, the vines benefit from those diurnal shifts, a perfect set of circumstances for growing its two prized grapes, Chardonnay and Pinot Noir.

Burgundy is divided into five subregions: Chablis, Côte de Nuits, Côte de Beaune, Côte Chalonnaise, and Mâconnais. Chablis is the northernmost, located closer to Champagne than to the rest of the region. The Côte d'Or ("Golden Slope"), which encompasses both the Côte de Nuits and the Côte de Beaune, is south of Chablis and is the most famous part of Burgundy, named both for the quality of wines produced as well as the color of the hillsides when the vine leaves turn yellow, orange, and red after harvest in the fall. Within the Côte d'Or, the Côte de Nuits ("Night Slope") is located in the north and named after the commercial center, Nuits-Saint-Georges, while the Côte de Beaune ("Slope of Beaune"), to the south, is

The Valois Dukes, who governed Burgundy from 1363 to 1477, played a critical role in turning the region's wine into a status symbol. Philip II the Bold, Duke of Burgundy, son of France's King John the Good, took a keen interest in Burgundian wines and understood their export value. He demanded growers cultivate Pinot Noir instead of any other red grape, and in 1395, he issued his infamous edict declaring Gamay to be "harmful to human health" and demanded growers uproot it.

named after the commercial center of Beaune. South of the Côte d'Or lies the Côte Chalonnaise, and south of that lies the Mâconnais, best known for Chardonnay in its hilly subregions. These two latter subregions are famous for delicious wines and terrific value, representing somewhat riper styles of Chardonnay.

I want to acknowledge that, yes, Burgundy is expensive and can be hard to navigate. I also want to assure you that there are now more excellent Burgundies than ever before, thanks to the region's global popularity and the fact that there are so many people making so many interesting wines in subregions that were considered off the grid until a decade ago. Also, while it will take time to be able to nail your Vosne-Romanée Premier Crus in a blind tasting (if that's something you care to do), there are some really beautiful aspects of this exceptional region that can make learning the basics encouraging. For starters, there is one white grape of supreme importance (Chardonnay) and one important red grape as well (Pinot Noir). The bedrock soil is always limestone mixed with clay. You can go down some very interesting rabbit holes about during which part of the Mesozoic Era (Triassic, Jurassic, Cretaceous) the limestone formed, and what type of ancient fossilized crustaceans helped create it, but just knowing you're dealing with limestone is a great starting place. And lastly, wine bottles labeled with "Domaine" mean that the winery farms, harvests, and vinifies its own grapes. These, except in rare cases (like Joseph Drouhin, which owns several vineyards and makes terrific wines), are just better. Wines labeled with "Maison" or anything other than the word "Domaine" mean the grapes were farmed by someone other than the winery. On a label, "Maison" indicates that the winery is a négociant. (See page 68 for some issues that even the best-intentioned négociants can face!)

CLASSIFICATION AND HIERARCHY:
GOOD, BETTER, AND BEST IN BURGUNDY AND BEYOND

In Burgundy, vineyards are ranked hierarchically beginning with the generic "Bourgogne" wines, representing more than 50 percent of all wines from the region. Above that, village-level wines represent about 30 percent of all vineyards in the region and earn the ranking because the vineyards are characteristic of that

143

"OLD WORLD" COUNTRIES AND REGIONS

particular village. Premier Cru vineyards, sometimes listed as 1er Cru, are exceptional vineyards representing about 10 percent of all Burgundy vineyard land. At the very top, the Grand Cru or "Great Growth" vineyards represent 2 percent of all land in Burgundy. They are prized for their exceptional aspect and slope, great drainage, a perfect balance of limestone and clay, and a reputation for excellence dating back centuries.

The vineyard hierarchy is often shown in a pyramid.

GRAND CRU VINEYARDS
The very best; these make up just 2 percent of all vineyards in Burgundy

PREMIER CRU VINEYARDS
Excellent; really great but not as great as Grand Cru; about 10 percent of all Burgundy vineyards are Premier Cru

VILLAGE-LEVEL
Notably good; vineyards are generally at the top (too rocky) or bottom (not rocky enough) of the slope but still distinctive enough to self-identify with the village they're a part of; about 30 percent of all Burgundy vineyards are classified this way

REGIONAL LEVEL
Last place; but still Burgundy

Burgundy's Geology

Millions of years ago, Burgundy looked like the Bahamas today, minus the cruise ships and motorboats. It was a big gentle lagoon full of crustaceans. Thousands of species of marine creatures like coral, oysters, and sea urchins lived, died, fossilized, and compressed over the course of about five hundred million years, becoming layers of limestone that remained underwater until Pangea began to separate about 175 million years ago. Millions of years later, the Alps emerged. Neither continent separation nor mountain creation is a gentle act, and as tectonic plates shifted and calcified marine layers thrust into the sky, the mountains of the Alps and subsequently the Pyrenees emerged. The Saône Valley, a natural fault line and now the backbone for Burgundy's storied slopes, was essentially collateral damage from mountain creation. The Saône Valley runs north to south and creates an almost continuous hillside of gentle slopes facing south-southeast. Distinct layers of limestone surfaced in each of these villages, giving rise to subtly different flavor profiles. Within each village, different vineyards show subtle differences, too. Burgundy's varying layers of limestone are all well known by the region's best growers, who have been raised with an acute awareness of, reverence for, and curiosity about the soils in their vineyards and la roche mère ("Mother Rock").

If you choose to go down the Burgundy rabbit hole, you'll invariably encounter information about different types of limestone: Jurassic, Bajocian, Bathonian, and the like. Don't let this confuse you! A lot of sources treat these as different kinds of limestone, whereas they are merely references to different time periods within the Jurassic Period during which the limestone was formed. For millions of years, the oceans were not conducive to animals forming crustaceous shells. This changed along with the ocean chemistry in the Cambrian Period about five hundred million years ago, when the ocean pH began to allow for shell production and an explosion of crustaceans emerged in the sea. Limestone, a calcium-carbonate-rich sedimentary rock, is in part made up of eroded and fossilized crustaceans whose skeletons consisted of calcium carbonate. Not all limestone is pure—calcium carbonate and impurities like silica can make limestone more porous and water-soluble, leading to erosion over time and transporting sediment to other parts of the land through groundwater. Many (and I am among them) believe this is what makes Burgundy's terroir so very special!

Grand Cru vineyards have been codified by everyone from monks to the modern French government. Its best examples are legendary; Burgundy's Grand Cru Chambertin vineyard produced Napoleon's favorite red wine. Today, these vineyards are so prized that they're recognized as their own individual appellations.

CHABLIS

This northernmost part of Burgundy, once regarded as too cool in climate to produce wines of consequence, is the cat's meow these days, for good reason. Kimmeridgian marl, plus probably magic, transforms Chardonnay grapes grown here into an elixir unlike any other. A fun piece of wine trivia: Chablis technically has only one Grand Cru vineyard, practically named Chablis Grand Cru. Within that one vineyard are seven subparcels, located next to one another, listed here in alphabetical order: Blanchot, Bougros, Les Clos, Grenouilles, Preuses, Valmur, and Vaudésir.

CÔTE D'OR

This is Burgundy's most famous real estate, made up of two smaller subregions, the Côte de Nuits and, to its south, the Côte de Beaune. It's hard to overstate the importance of this subregion—many wine lovers save up for years in order to make a pilgrimage to these hallowed vineyards, and it remains (I would argue) a spiritual place. The Côte de Nuits is home to the most famous Pinot Noir in the world, made in villages that include Gevrey-Chambertin, Morey-Saint-Denis, Chambolle-Musigny, Nuits-Saint-Georges, and Vosne-Romanée. The Côte de Beaune, just south, is home to the world's most storied villages for Chardonnay, including Meursault, Puligny-Montrachet, and Chassagne-Montrachet. Today, the Côte d'Or's vineyards are a UNESCO World Heritage Site and represent the origin of an appreciation for single-vineyard wines.

146

MERSAULT

The village of Meursault has a particularly large number of lieux-dits (literally "named places"), classified as over-performing, village-level parcels. Despite its reputation as one of Burgundy's greatest villages for white wines, is notably short on Grand Crus (it doesn't have any). Current vineyard classifications date back to 1955, but older maps show that Les Perrières, a vineyard in Meursault that borders Chassagne-Montrachet, used to be ranked as Grand Cru and considered one of the greatest vineyards in the land. When I asked a few growers why this changed, their answer was simple: growers in the villages of Chassagne and Puligny were wealthier and had political ties when the rankings were cast. Growers in Meursault did not, and they missed out on Grand Cru classifications because of it.

CÔTE CHALONNAISE

Burgundy's Côte Chalonnaise extends just south of the Côte de Beaune and benefits from limestone soil, but not the protection or uniformity of a continuous slope. As a result, vineyards are planted among different rolling hills, and even though the location is more southerly, temperatures are cooler because of frequent winds. There are five AOPs located here, including Bouzeron, the one appellation in France that makes wines from exclusively 100 percent Aligoté. Rully, Mercurey, and Givry all make white and red wines from Chardonnay and Pinot Noir, and Montagny makes wines from 100 percent Chardonnay.

MÂCONNAIS

This region of Burgundy, south of the Côte d'Or and north of Beaujolais, is full of beautiful rolling hills covered in vines, pastures, farmhouses, and fields planted with things other than just grapevines. The Mâconnais is named after the town of Mâcon. The sun shines a bit warmer here, and the region is known for some of the world's most delicious *and* best-value Chardonnay, somewhat riper, rounder, and juicier than Chardonnay from its northerly neighbors in the Côte d'Or and Chablis. Happily, it's also usually less of an investment—and yet the wines can age beautifully!

On labels, look for the following, in order of most general to most specific in the region:

1. Mâcon AOP

2. Mâcon-Villages

3. Mâcon-Vergisson (Mâcon + village name ; for red wine, Mâcon + village name is made from 100 percent Gamay, not Pinot Noir)

4. Pouilly-Fuissé (the name of a specific subregion within the Mâconnais)

JURA

Nestled close to the Swiss border in a cool, humid, and subalpine climate, the Jura is a small appellation producing spectacular wines that are sort of like spunkier, less-polished relatives of its more westerly neighbor with similar grapes and soil (Burgundy). The Jura is a beautiful region of rolling limestone hills that boasts a mother rock of blue marl that predates the limestone in Burgundy. Many forward-thinking and traditional wine growers are pushing boundaries and making some of France's most interesting wines here. Trousseau and Poulsard are the region's two ancient indigenous red varieties; Trousseau is a bit darker in color and more structured than Poulsard. Oxidative "biological" wines aged under a thin veil of yeast called le voile will exhibit distinct, almost sherry-like aromas that are uncannily well-suited to a variety of foods. This region is home to vin jaune, aka yellow wine (see page 53).

BEAUJOLAIS

Beaujolais is technically part of Burgundy, and some of the region overlaps with the Mâconnais (Burgundy's southerly subregion), but most of it is located south, closer to the Rhône Valley. Beaujolais is famous for its granite soil and the red grape Gamay. More and more, great Gamays from Beaujolais are beginning to get the credit they deserve. The most long-lived of the world's benchmark examples of Beaujolais (and 50 percent of all Gamay vines worldwide) come from one of ten villages located in the north, which are somewhat confusingly called Cru (which, in this case, means "village"). These ten Beaujolais Crus run north to south and benefit from soil like decomposing granite, pink granite, and fossilized ash from extinct volcanos, which all produce markedly different and very yummy expressions of Gamay. The Crus, listed north to south, are as follows: Saint-Amour, Juliénas, Chénas, Moulin-à-Vent, Fleurie, Chiroubles, Morgon, Régnié, Côte de Brouilly, and Brouilly. Reputationally the lightest examples, which are best in their youth, tend to come from the villages Brouilly, Régnié, and Chiroubles. Crus that are terrific at any age but have more structure (and develop in bottle a bit longer) include Saint-Amour, Fleurie, and Chénas. Crus that produce the longest-lived wines are the Côte de Brouilly, Juliénas, Moulin-à-Vent, and Morgon. The latter two represent the longest-aged wines, benefiting from pink granite (Moulin-à-Vent) and decomposing granite soils (Morgon).

Beaujolais is also known for Beaujolais Nouveau, a cheap and cheerful wine made by a process called carbonic maceration, where harvested grapes are kept in a tank filled with CO_2 (and no oxygen), and fermentation happens inside of the grape berries, producing a juicy wine with soft, barely perceptible tannins from all that intracellular fermentation. That wine is released on the third Thursday of November each year. This style was created to generate immediate cash flow for wineries who opted to produce it—and while it long had a reputation for low quality, quality-oriented producers like Domaine Lapierre now release versions, partly in irony, but partly because it can be very good! One note—these wines are meant to be opened and drunk immediately. Don't try to keep half of a bottle in the fridge for the next day. I don't know the chemistry behind it, but day-old Beaujolais Nouveau wines invariably taste like dill pickle brine.

BORDEAUX

When I visited Bordeaux for the first time, I was struck by the different histories, soil types, and philosophies between the Left and Right Banks—and I wish this information had been clearer in books I'd read before visiting.

BORDEAUX'S LEFT BANK

The Left Bank, encompassing the Médoc (to the far north) and the Haut-Médoc (directly south of Médoc, including the more famous appellations Saint-Estèphe, Pauillac, Saint-Julien, and Margaux), lies along the west or "left" bank of the Gironde estuary. It is a relatively new wine region that dates back to the mid-seventeenth century, when the Dutch drained what had previously been a salt swamp in order to have vines under their control. (They dominated rivers and seas but had to rely on other countries for wine, which annoyed them.) After draining the swamp, they uncovered gravel mounds in Bordeaux, planted vineyards, and found they could produce wines that were competitive with those made across the Gironde River, aka the Right Bank, especially with good marketing. The Left Bank is associated with Cabernet Sauvignon, which can make extraordinary wines on gravel soil if they're able to ripen (historically a challenge in the area). Unripe Cabernet Sauvignon has a green, stemmy taste, hence the long, now-prohibited tradition of shipping Syrah up the river and blending it with the Cabernet in what is known as Hermitagé.

Even though it emerged centuries after the Right Bank, the Left Bank's proximity to the Atlantic Ocean made it an important commercial center. By the mid-nineteenth century, a merchant class formed in the industry that was able to promote and profit from the Left Bank's wines on a commercial scale. Even today, the vibe in the Médoc is "business first," with well-known estates investing in theatrical architecture and plenty of new winemaking technology. (I remember visiting Château Cos d'Estournel and feeling I was at the old NoMad Hotel in Manhattan, before learning the same person, the famed maximalist architect and designer Jacques Garcia, had designed both.)

The most famous villages on the Left Bank known for producing dry and mainly red wines, listed from north to south, include Saint-Estèphe, Pauillac, Saint-Julien, Margaux, Pessac-Léognan (to a lesser extent), and Graves. Far south of those are the trio of appellations: Cérons (*SEH-rone*), Barsac (*BAR-sack*), and Sauternes (*sow-TERN*), all famous for making sweet and semi-sweet wines from botrytized white grapes.

One estate that has always stood out to me for its commitment to quality and sustainability is Château Pontet-Canet in Pauillac, the first classified estate in the Médoc to seek organic and biodynamic certification. I find these wines inspiring and delicious, and a fraction of the price of First Growth Bordeaux. I am also partial to Château Lynch-Bages, another overperforming Fifth Growth located in Pauillac. Diana Snowden Seysses turned me on to this wine at a blind tasting over dinner during the harvest at Domaine Dujac in 2009. We drank the 1986 vintage and it was astoundingly delicious!

The famous 1855 Bordeaux Classification of the Médoc, which ranks châteaux from First Growth through Fifth Growth, was determined by brokers (middlemen) ranking wineries based on their market prices at the time of classification, just before the 1855 World's Fair held in Paris. In theory, the First Growths are the best, with Second, Third, Fourth, and Fifth Growths following in a meritocratic fashion. In reality, these rankings are far more reflective of the market value in 1855.

BORDEAUX'S RIGHT BANK

The Right Bank (farther east and to the "right" of the Gironde River on a map) has a much longer viticultural history than the Left Bank, dating back to the Romans in the fourth century BCE. Merlot is the star here, grown primarily on clay soils. It ripens later than Cabernet Sauvignon and is almost always an important ingredient in even the best Cabernet Sauvignon–based wines. The Right Bank of Bordeaux did not classify its wines in 1855 as the Médoc (Left Bank) did, and its greatest wines are not ranked. (Saint-Émilion, whose wines were supposedly favored by King Louis XIV, releases classifications of it's own wines every ten years, but these have never entered the zeitgeist the way the 1855 Médoc classification has). The two most famous villages on the Right Bank are Saint-Émilion and Pomerol. Saint-Émilion is celebrated for its limestone soil (Calcaires à Asteries) and for growing world-renowned Cabernet Franc. Pomerol is revered for Merlot grown on clay soils, producing wines that are rich, velvety, and sumptuous. The most iconic wines include Pétrus (Pomerol), Château Lafleur (Pomerol), and Château Cheval Blanc (Saint-Émilion), the latter known for exalting Cabernet Franc as the star.

ENTRE-DEUX-MERS

Between the Gironde and Dordogne Rivers, in what is neither the Left nor the Right Bank, is an area known as Entre-Deux-Mers, or "Between Two Seas," referring to its location between Bordeaux's two main rivers. Dry white wines made from blends of primarily Sémillon and Sauvignon Blanc are the best known from the region, as are sweet and off-dry wines made from the same grapes.

RHÔNE VALLEY

Named for the Rhône Glacier, which originated in Switzerland and then carved its way to France and all the way to the Mediterranean during the last Ice Age, the Rhône Valley includes two parts—Northern and Southern. Like Bordeaux, these two very different regions have varying climates, soils, and grapes. They share a river and therefore a name, but that's about it.

The Northern Rhône wine region begins just south of Beaujolais and is known for savory red wines made of Syrah, and, to a lesser extent, textural white wines from Marsanne and Roussanne (usually blended together) and occasionally white wines made from the perfumy grape Viognier. Broadly speaking, the soil in the region primarily comes from metamorphic rock, with gneiss, schist, and granite as the main players. And while white grapes are grown and often even included in the red wines (see page 152), the Northern Rhône is best known for red wines made from Syrah grapes. In the Northern Rhône, Syrah can be inky and almost black, especially in areas like Cornas. Syrah from the Northern Rhône reminds me of pastrami, leather, blackberries, and green tobacco. While the grape is grown throughout the world, I've always found it to be more interesting when grown here than anywhere else.

The region's major subregions, listed from north to south, include the Côte-Rôtie ("Roasted Slope"), home to cool-climate Syrah grown on mainly gneiss and schist on the western slopes of the Rhône River. Just south lies Condrieu, an appellation that specializes in perfumy, viscous Viognier and makes wines from exclusively that grape. Within Condrieu is a tiny and historic appellation called, confusingly, Château-Grillet. The entire appellation is an ampitheater-shaped monopole (a vineyard with a single owner) made by a winery that is also called Château Grillet.

South of Côte-Rôtie and Condrieu is the largest appellation, Saint-Joseph, which snakes from north to south alongside the western bank of the Rhône River and grows mainly Syrah on granite soil. When the appellation was created in 1956, there were six villages; today the AOP has expanded to twenty-six! While most wines from the region are red, white wines (from Marsanne and Rousanne, often blended together) are available, too, and they can be very good. Saint-Joseph is also responsible for some truly beautiful everyday wines that cost between twenty to sixty dollars a bottle and can age for a few years in the cellar, getting more delicious and complex all the while. Syrah from Saint-Joseph tends to have some grippy tannins, thanks to that beautiful granite soil. Four terrific producers all focus on vineyards in the south, surrounding the town of Mauves—Domaine Jean-Louis Chave (also famous for Hermitage bottlings), Pierre Gonon, Coursodon, and Domaine Bernard Gripa are all based here. It turns out that Mauves's fame predated even them, however—Victor Hugo references the good wine of Mauves in *Les Misérables*! Two producers I have long admired in the northern part of the appellation include Domaine Monier Perréol and Yves Cuilleron.

Across the river from the southern end of Saint-Joseph lies the hill (well, technically, two hills) of Hermitage. It's a marvel of a wine region situated on the part of the Rhône River that turns east before resuming its southward flow,

Sometimes white grapes are added to red wine—a common practice in the northern parts of the Northern Rhône, where a percentage of white grapes are allowed in the reds. Back in the days when getting grapes to fully ripen was a challenge, adding some white grapes helped—they ripened earlier and helped boost sugar and alcohol levels, almost as a form of grape-based chaptalization. Alcohol is a purveyor of flavor, almost like fat in cuisine.

creating two giant hills with south-facing slopes which have miraculously diverse soil (thanks to a bunch of river deposits from the last Ice Age). The slopes are sun-drenched and therefore historically prized during a time when it was particularly hard to ripen grapes. The hills tower above the river below, and vines are often head-trained as the slopes are far too steep to plant in rows. If you have a chance to walk among the vines, you will feel a palpable reverence for history. Nearby, Crozes-Hermitage profits from its proximity to the hill of Hermitage and co-opted its more famous neighbor's name. Domaine Alain Graillot makes gorgeous wines, as does Domaine Jean-Louis Chave, but a vast majority of Crozes-Hermitage is as flat as a wheatfield and has nothing in common with the terroir that makes Hermitage so special!

Last but not least when it comes to red-wine-only subregions is Cornas, which makes wines from exclusively Syrah and sits on the west bank of the Rhône River south of Saint-Joseph. (Saint-Peray is technically the last subregion of Northern Rhône and makes lovely, but far less well known, white wines from Marsanne and Rousanne.) The east-facing vines get plenty of warmth because the location is so far south. Wines from Cornas tend to be the darkest and meatiest of the Northern Rhône, thanks largely in part to the decomposing granite soil known as "gore," mixed with small amounts of limestone and sand. Top producers include Noël Verset (who retired in 2006, but if you stumble upon a bottle, it's quite special), Domaine Clape, Thierry Allemand, Franck Balthazar, and Domaine Vincent Paris.

SOUTHERN RHÔNE

The Southern Rhône region, like the Languedoc, is known for producing a lot of wine of varying quality. Châteauneuf-du-Pape is the region's most famous appellation, with a history dating back to the fourteenth century when the papacy moved from Rome to Avignon. Today, the Southern Rhône's largest export is from Côtes du Rhône, usually red and Grenache-based, with support from Syrah, Mourvèdre, Carignan, and some other minor players.

Châteauneuf-du-Pape famously allows thirteen grapes in its blend, and there are some profound wines in the appellation, including Château de Beaucastel and Domaine du Pégau. The appellation's most famous winery is Château Rayas, owned by the Reynaud family since the 1880s. Rather than succumbing to the urge to blend grapes together, the Reynaud family has focused on Grenache for their grand vin, Château Rayas, as well as a bottling called Pignan, adding some Cinsault and Syrah to the blend. In some vintages the domaine will release a Fonsalette made from 100 percent Syrah grapes, and it's extraordinary! We carried it at Eleven Madison Park, which I bring up here as a note that one terrific reason to go to great restaurants is that winemakers are happy to send their best wines there, so you'll find—and be able to drink!—bottles that don't often appear elsewhere.

White wines are primarily made from Grenache Blanc, Clairette, Bourboulenc, and Rousanne grapes. Tavel, which by law can produce only toothsome rosé wines, is equally notable as the place where phylloxera entered France.

Grenache, the most prominent grape of the Southern Rhône, is juicy and ripe, like a strawberry in July. Grenache (or Garnacha in Spain, where it is originally from) thrives in warm climates and tends to produce fruit-forward wines with high alcohol levels and low tannins. Grenache is all about Provençal herbs, sweet tobacco, and ripe strawberries.

PROVENCE AND LANGUEDOC

Azure sea and skies and pristine weather make southern France one of the most beautiful places on earth. It also has a rich viticultural history and a diverse array of permitted grape varieties. Throughout various points of history, the ancient Greeks, Gauls, and Romans, as well as the emperor Charlemagne and the country of Sardinia, controlled this part of the world. One subappellation of Provence, Palette, allows up to nearly twenty grapes! Southern France's one climatic worry involves the mistral, an extremely strong wind that sweeps through the region. This leads to deeper roots in the vines (and thus more pronounced minerality in the wines), as well as thicker skins (which creates stronger tannins).

The star of Provence's wine scene, aside from rosé, is red from the Bandol appellation, made from the Mourvèdre grape. I remember drinking a bottle of old Domaine Tempier with Daniel Boulud and Robert after the James Beard Awards one night in the early 2000s—Daniel described Mourvèdre (aptly) as tasting like "blood and guts in the sun." The wines are gamey and sanguine, rich and intense—yet, thanks to the limestone-clay soils and cool nights from the mistral, the wines have lift and brightness and are quite special, particularly with bottle age.

Separately, the Languedoc is one of France's largest production regions, with a climate that makes vine growing almost too easy. For much of its history, the Languedoc has been associated with overproduction. One star producer that shines above the rest is Laurent Vaillé of Domaine de la Grange des Pères, who worked at Domaine de Trévallon in Provence, Domaine Coche-Dury in Burgundy, and Domaine Jean-Louis Chave in the Northern Rhône before deciding to purchase a plot of land he deemed worthy of his life's work. He makes white wines from Marsanne, Rousanne, and Chardonnay, as well as red wines from a blend of Syrah, Mourvèdre, and Cabernet Sauvignon.

GERMANY

WHERE RIESLING REIGNS
(AND THE PINOT NOIR IS PRETTY GREAT, TOO)

Buckle up! So many German wines are pure, profound, and beautiful. Their wine label terminology, however, is overly complex. I go into some detail below, but after a few deep breaths and further reflection (on my end), I remembered that this book is not trying to educate you on the intricacies of German wine law and its seemingly interminable amendments, the most recent of which will go into effect in 2026. (There are plenty of websites that do exactly that!) Instead, it is meant to help navigate you through the noise and save you some time for things that matter in life, like drinking great bottles of wine with people you like.

Geographically, Germany is located north of other wine-producing countries, and prior to global warming, grapes struggled to ripen. I share this because Germany is full of grape varieties that we're not going to spend time discussing, like Müller-Thurgau, which was created in the late nineteenth century by a Swiss botanist at a research institute for the express purpose of making grape growing easier for growers who lived in a cold climate. Instead, we're going to focus on the grapes I believe Germany does best—namely Riesling and Spätburgunder (aka Pinot Noir).

Like so many European countries, Germany's viticultural history arrived with the Romans, who crossed the Alps and continued eastward more than two thousand years ago. Winemaking prospered under Charlemagne in the eighth century, and orders of Cistercian monks from Burgundy set up monasteries in Germany, growing both Riesling and Pinot Noir through the Middle Ages. German wine during the seventeenth and eighteenth centuries had peaks and valleys, hindered by wars and a string of extremely cold winters, but the nineteenth century was a bit of a golden age when Napoleon liquidated church holdings after the French Revolution and estates were privatized.

German Rieslings (including sparkling Sekts) were some of the most revered in all of Europe in the late 1800s, until a series of setbacks—phylloxera, oidium, a pause in the American export market during Prohibition, and two world

GERMANY
WINE REGIONS

GERMANY

Berlin

Frankfurt

Stuttgart

Munich

Rhine

Mittelrhein

Ahr

Rheingau

Frankfurt

Main

Mosel

Mosel

Franken

Nahe

Nahe

Rheinhessen

Hessiche- Bergstrasse

Pfalz

FRANCE

Rhine

Stuttgart

Danube

Neckar

Baden

Lake Constance

Württemberg

SWITZERLAND

AUSTRIA

wars—halted the momentum. While World War I didn't affect the actual vineyards as much as World War II did, it took workers out of the vineyards and off to battlefields. World War II's impact was grave for several reasons. Notably, the Nazi government drove out the Jewish population, who accounted for approximately 70 percent of the wine trade. Additionally, the Nazi government introduced mechanized farming, chemical vineyard treatments, and a focus on quantity over quality in wine, looking at viticulture as a revenue stream instead of an art form. Happily, that has changed, and many of Germany's great producers today are crafting wines that respect terroir, reinvigorating vines abandoned after phylloxera, and bucking the rules put in place during the Nazi reign, all in the pursuit of excellence.

GERMAN WINE LAW TODAY:
A VERY BRIEF OVERVIEW

Germany has thirteen designated wine regions that are collectively known as the Anbaugebiete (*ahn-BOW-guh-beet-uh*), which were formally created with the German Wine Law of 1971 for global export, a response to Europe's postwar attempts to standardize labeling and regulation. These thirteen regions are primarily located in western Germany and located along rivers, which historically served as the mode of transport for finished wines (and everything else). Rivers have the added benefit of moderating Germany's cool temperatures and helping grapes in nearby vineyards ripen. Germany's Anbaugebiete are equivalent to French AOPs, Italian DOPs, and Austrian DACs.

Tragically, the German Wine Law of 1971 reduced the number of individually recognized vineyard sites, shrinking the total from about thirty thousand to fewer than three thousand. The law essentially smooshed great vineyards in with bad ones and co-opted the better-known names, with the intention of using the famous and recognizable wines to market crappier ones throughout the world. As you can imagine, this fueled customer confusion (which it had intended to minimize) as well as vitriol and angst from quality-oriented German wine growers who had long prioritized the magic of individual vineyard sites. Additionally, the German Wine Law of 1971 created the Prädikat system, which confoundingly ranked a wine's quality based on the level of sugar in the grapes at harvest. Thanks to a 2021 amendment (which will go into effect in January 2026), a new and more consumer-friendly system will prioritize terroir over sugar levels as a defining factor for quality.

The Prädikat System, or Prädikatswein

While the terms below are being phased out of the German Qualitätswein official categorization, they're still likely to show up on labels, and it's useful to have a handle on what they mean. Specific Oechsle scale levels are mandated for each of the categories below. (The Oechsle scale is Germany's system for measuring sugar density, or the weight of grape must, and is abbreviated as °Oe.) The low end of the range applies to Riesling from Germany's coolest Anbaugebiete, the Mosel. In the list below, the percentage that follows the Oechsle scale range corresponds to the amount of potential alcohol, should all of those sugars hypothetically ferment.

Kabinett	These wines are a little bit sweet and qualified by a range of 70 to 85 °Oe and 9.1 to 10.9 percent potential ABV.
Spätlese	These wines are notably sweeter than Kabinetts but still very balanced and not too heavy in palate weight. They range from 76 to 95 °Oe and 10 to 12.2 percent potential ABV.
Auslese	These are dessert wines, made from late-harvest grapes. They range from 83 to 105 °Oe and 11.1 to 13.8 percent potential ABV.
Beerenauslese	These are dessert wines made from individually botrytized grapes (and therefore a different and complex flavor profile, with notes of ginger, saffron, and chamomile, thanks to the botrytis influence). These wines range from 110 to 128 °Oe and 15.3 to 18.1 percent potential ABV.
Trockenbeer-enauslese	Made from individually hand-picked grapes that are dried but *not* botrytized. These grapes are picked even later than the Beerenauslese and are some of the sweetest (yet also marvelously balanced) in the world. These range from 150 to 154 °Oe and 21.5 to 21.9 percent ABV.
Eiswein	This is made from grapes that have shriveled on the vine, *not* succumbed to botrytis, and that are harvested the night—or early morning—of the first frost. The grapes are then pressed immediately, extracting whatever juice remains while the frozen water inside remains behind and is discarded after pressing. These wines are bright, vibrant, and sweet but not cloying, with a range of 110 to 128 °Oe and 15.3 to 18.1 percent ABV.

The VDP

The Verband Deutscher Prädikatsweingüter, or VDP, is an invitation-only organization made up of quality-minded growers from within each of the thirteen Anbaugebiete. These growers are committed to preserving traditional grape varieties as well as terroir. They've taken inspiration from Burgundy in celebrating and prioritizing historically great vineyard sites and appear to have influenced the 2021 amendment to the German Wine Law of 1971. Wines considered the highest quality by the VDP are labeled "VDP.Grosse Lage®."

Wines made by VDP members are indicated by the VDP logo of an eagle surrounding a bunch of grapes, printed on the bottle's foil capsule. As of 2011, VDP members can produce wines in four different categories, emulating Burgundy. From lowest to highest in quality, the VDP system includes a regional level (Gutswein), a village level (Ortswein), premier cru vineyards (Erste Lage), and grand cru vineyards (Grosse Lage).

More German Wine Label Terminology!

Trocken
The term means "dry"—and if you see it on a label of German Riesling, you can expect a dry, generally very zippy glass of wine with no perceptible residual sugar. By law, this means the wine has a maximum of 9 g/l of residual sugar. That might seem like a lot until you consider that Coca-Cola has 108 g/l of residual sugar! Sweetness is a function of perception, and Riesling has so much acidity that this small amount of residual sugar is essentially canceled out.

Grosse Lage
This term means "great site" and is a VDP-trademarked term reserved only for sites deemed the very best in all of Germany. To make this assessment, the VDP considers everything from old vineyard maps to current vineyard potential. According to its website, "Wines from these vineyards shine through their uniqueness and distinctiveness."

Erste Lage
The VDP's name for premier cru sites, ranked just beneath Grosse Lage.

Erstes Gewächs
A term originally used to denote wines from the highest-quality vineyards vinified in a dry style. The term is now used by Rheingau growers in a group called Charta (who have decided not to participate in the VDP). Their identifying logo is a stylized pair of Roman arches.

Grosses Gewächs
The term was created and used by the VDP for wines made from the highest-quality vineyards vinified in a dry style. This term was inspired by the term "Erstes Gewächs." (There is some long-standing beef between some Rheingau growers and VDP growers, and it has manifested in two different terms for the same thing.)

In addition to the Prädikatswein category, the other three are below:

- **QUALITÄTSWEIN:** This category covers quality wine with a protected designation of origin, which must be one of the thirteen Anbaugebiete. Quality level is based on specific geographic origin, with single vineyards being the most esteemed.

- **LANDWEIN:** This category requires that 85 percent of grapes must come from a Landwein region, which are different from the Anbaugebiete. These regions will be named on the label (but we're not going to list them here because these wines don't typically account for the majority of German wine in the US market).

- **DEUTSCHER LANDWEIN:** This is Germany's table wine category, which is the least quality-oriented and least expensive.

In this chapter, we'll focus on five of the thirteen regions—the Rheingau, the Mosel, Rheinhessen, Baden, and Ahr. However, there are worthy wines from all of Germany's regions. The Nahe makes plenty of stunning Rieslings ranging from bone-dry to dessert-sweet styles (Dönnhoff and Emrich-Schönleber are two standout producers). The Pfalz, right across the French border, is almost like an extension of Alsace and produces excellent Weissburgunder (Pinot Blanc), Grauburgunder (Pinot Gris), and Spätburgunder (Pinot Noir). Looking at a map, you'll notice that almost all of Germany's wine regions are in the southwestern portion, near Switzerland and France.

THE RHEINGAU

Riesling is by far Germany's most significant grape, with Cistercian monks at the famous and influential monastery Kloster Eberbach praising it back in the fifteenth century. Over 40 percent of the world's Riesling vines are planted in Germany today. Historically famous regions include the Rheingau, with its red slate soil and rare south-facing slopes leading down to the famed Rhine River, where most experts believe the grape originated. Depending on who's calling the shots, Rheingau Riesling can be made bone-dry or sweet, vinified in large oak barrels or stainless steel, and fermented with ambient (natural) or inoculated (lab created) yeasts.

In the Rheingau particularly, 80 percent of all vineyards are planted with Riesling, including all of the best vineyard sites. Several famous wineries still bear names referring to prior eras when monestaries and nobility held power and sway. (The word *schloss* refers to a castle and the word *kloster* to a monastery, and plenty of wineries begin with one of these two names.) This region is also home to Germany's first Spätlese and Auslese wines dating back to the 1700s, made from late-harvest botrytized grapes that transformed in humid vineyards nearest the Rhine River.

In 1984, a group of Rheingau growers, who were annoyed with the German Wine Law of 1971 but disinterested in joining the VDP, created their own organization, Charta, which has a similar aim—high quality and historic vineyard sites—but a different name for the best parcels (it calls its top vineyards Erstes Gewächs and has a different identifying logo of two Roman arches). Think of Charta as the Rheingau's proprietary take on the VDP, with a different name and logo.

THE MOSEL

One of the world's most northerly growing regions (located above the 50th parallel north) and one of the coolest, the Mosel is named for the river flowing off of the Rhine that snakes through steep hilly slopes. It's one of the world's most dramatic, picturesque regions. Vineyard slopes are commonly 50 to 60 degrees, with some as precipitous as 65 degrees. As a result, single-vineyard sites in the Mosel are incredibly challenging for vine growers and were some of the first abandoned when times got tough, such as during the phylloxera disaster of the late nineteenth century and both world wars. Viticulture is possible here on account of the dark blue-gray Devonian slate, which collects heat during the day and releases it back onto the vines at night, as well as the south-facing steep slopes on the north banks of the river, which reflect light off of the river and back onto the vines like a mirror.

The Mosel's viticultural history is Germany's longest, dating back to the Romans in 16 BCE. Riesling remains the region's most important grape. Historically, the Mosel was able to produce light and elegant sweet and off-dry wines, in contrast to the fuller-bodied sweet wines of the Rheingau, even before the invention of sterile filtration. The delicate, off-dry Kabinett style was invented here. Wineries Egon Müller, Weiser-Künstler, Joh. Jos. Prüm, Peter Lauer, and von Schubert are among my longtime favorites.

RHEINHESSEN

While Germany's most famous soil is slate, the Rheinhessen has two notable vineyard areas—the Roter Hang, a narrow slope of red sandstone, and a subregion called Wonnegau, which has a stretch of gentle limestone slopes made famous by two producers, Klaus-Peter Keller and Philipp Wittmann, who make exceptional wines there today. The grapes Müller-Thurgau and Sylvaner still have a large presence in the region, making some easy-drinking, if less serious wines.

BADEN AND AHR

The Ahr (one of Germany's smallest regions and the most northerly) and Baden (the third-largest and most southerly) are the country's two superstars for Spätburgunder, or Pinot Noir. Vineyards in the Ahr, like those in the Mosel, benefit from the dark blue-gray Devonian slate that absorbs sunlight and heat during the day and reflects it back to the vines at night. The Ahr also profits from sheltered slopes and a turn in the jet stream that redirects cold air elsewhere. In the Ahr, Pinot Noir accounts for 70 percent of the vineyards. The winery Meyer-Näkel is often credited with the region's increasingly strong Pinot Noir reputation. After lackluster reviews in the 1980s, it took cues from Burgundy and began fermenting to dryness and vinifying the wines in French oak.

Baden has a long history of Pinot Noir, dating back to 884 CE, when the grape accompanied Emperor Charles the Fat (great-grandson of Charlemagne) from what is now France to what is now Germany. Baden runs parallel to both Alsace and the Pfalz (although farther east) all the way to the Black Forest. The region benefits from the same rain shadow effect and the same mosaic of soils as in Alsace and Pfalz, including granite, limestone, and volcanic soil, along with those trademark warm, sunny days. In Baden, the challenge is not letting Pinot Noirs become *too* ripe—an opposite obstacle from every other region in Germany. With ripeness not an issue, many wineries ferment their Spätburgunder whole-cluster. Among the superstars of the region, Enderle & Moll is a standout. It crafts some of my favorite Pinot Noir wines in the world, full stop, and long before it was trendy, it paid attention to every detail, going to the trouble of sourcing the highest-quality used barrels from wineries like Domaine de la Romanée-Conti and Domaine Dujac.

ITALY

It's impossible to pigeonhole Italy's wine identity, which is also so much of its charm. Despite the country's incredibly relevant role in Western history, it was not technically a unified nation until 1861 (the same year a young United States embroiled itself in civil war). Trentino-Alto Adige, north of Veneto, did not join Italy until 1919, after World War I, and even today five of Italy's twenty regions are considered autonomous, not including the Vatican (which is its own independent nation within the capital city of Rome). Viticulturally, this strong independent streak has benefited Italy, a country that has in many cases resisted the urge to uproot indigenous vines in favor of ubiquitous grapes like Cabernet Sauvignon and Chardonnay. While all twenty of Italy's regions produce terrific wines well worth exploring, we're going to focus on those I think you're most likely to encounter and enjoy. There are so many great books dedicated to the wines of Italy. I've recommended some favorites in the appendix (see page 236).

Italy followed France's AOP designation, creating its own DOC system (Denominazione d'Origine Controllata, which means "Denomination of Controlled Origin") in 1963. The organization rebranded as DOP (Denominazione d'Origine Protetta or "Protected Designation of Origin") in 2009. The system is meant to protect quality and set consumer expectations for foods and wines labeled with DOP certification. Italy added an upgrade, the DOCG system ("G" for guaranteed), in which wine requires a laboratory assessment as well as a panel tasting. DOCG on a label indicates that there is an intention for the highest quality. The first five DOCGs to be awarded, beginning in 1980, include Barolo, Barbaresco, Brunello di Montalcino, Chianti, and Vino Nobile di Montepulciano (the latter being the first to receive DOCG status, having applied as early as 1969).

Geographically, Italy benefits from a varied landscape almost entirely covered in hills or mountains, and almost always relatively close to a coastline. Essentially, the country is perfect for organic viticulture. The ancient Greeks recognized this, calling southern Italy "Oenotria," or "land of wine." The Alps and Dolomites line the northern part of the country, crossing east to west, and the Apennine Mountains run up its center, north to south, like a spine. Italy has become a world leader in strict agricultural standards, limiting glyphosate (the herbicide Roundup, a neurotoxin and carcinogen; see page 95) since 2016. Italy is also at the forefront of viticultural sustainability. The Italian-based organization Equalitas has created a sustainability certification called 3E, representing environmental, economic, and ethical sustainability. Hopefully this sort of certification is the first of many worldwide.

WINE AND TAXES

Italy's five autonomous states include Friuli-Venezia Giulia, Trentino-Alto Adige, and Valle d'Aosta (which are all located at the northern border), along with the islands Sardinia and Sicily. While these regions still function as part of Italy, they are allowed to hold on to a larger percentage of taxes—60 percent instead of the usual 20 percent (or in Sardinia's case, 100 percent) in order to preserve each region's unique cultural heritage, of which wine is certainly a part!

ITALY
WINE REGIONS
SWITZERLAND
AUSTRIA
SLOVENIA
CROATIA
Trentino-Alto Adige
Friuli-Venezia Giulia
Lombardy
Veneto
Valle D'Aosta
Adige
Venice
Po
Piedmont
Genoa
Liguria
Emilia-Romagna
Florence
Arno
Le Marche
Tuscany
Umbria
ADRIATIC SEA
MEDITERRANEAN SEA
Abruzzo
Lazio
Molise
Rome
Campania
Puglia
Naples
MOUNT VESUVIUS
Basilicata
Sardegna
SARDINIA
Cagliari
Calabria
TYRRHENIAN SEA
Palermo
MOUNT ETNA
SICILY
IONIAN SEA

FRIULI-VENEZIA GIULIA
AND TRENTINO-ALTO ADIGE

Friuli-Venezia Giulia, in northeast Italy, has, at various times, belonged to Austria, Hungary, Turkey, and the Roman Empire. Today, it borders Slovenia, where the region continues under the Slovenian name Goriška Brda (*gore-EESH-ka BUR-dah*). Orange wines owe their recent popularity to a resurgence of the style in this neighborhood, from winemakers like Movia (from Slovenia) and Radikon (from Italy).

Lodged between a band of two Alps mountain chains and the Adriatic Sea, Friuli-Venezia Giulia's climate is as varied as its cultural history. The region's best wines are generally grown in the eastern hills, the Colli Orientali. There, high elevation provides cool temperatures, and winds from the mountains blow away humidity and prevent rot. The Adriatic Sea moderates the climate, ensuring that vineyards are never too hot or too cold. Venica & Venica, Borgo del Tiglio, Miani, and Vignai da Duline are some of my favorite producers anywhere, and they are all based in Friuli-Venezia Giulia. The region's full grape lineups are terrific, from Friulano (a white grape that, as you probably guessed, celebrates the region and tastes of golden apples and sunflowers) to Sauvignon Blanc, Pinot Grigio, and even red grapes Merlot and Schioppettino (*skio-pet-TEE-noh*). Sommelier and restaurateur Bobby Stuckey is almost single-handedly responsible for introducing most of America to these wines and getting people excited about them. His downloadable wine list at Frasca Food and Wine in Boulder, Colorado, is a terrific place to look if you want to explore this region more.

Trentino-Alto Adige lies west of Friuli-Venezia Giulia and away from the coast, but its brisk alpine climate is similar. In this region, winemakers like Elisabetta Foradori are giving "it-girl" status to grape varieties previously headed for extinction, like the red grape Teroldego and white grape Nosiola, the latter of which is likely named for (and tastes of) hazelnuts. This region is also home to the charming and delicious (but not very sexily named) red grape Lagrein (*luh-GRINE*).

VENETO

Located in northeastern Italy, just south of Friuli-Venezia Giulia along the Adriatic Sea, Veneto produces more wine than any other region in the country. Pinot Grigio is the workhorse here. It's often overcropped and can taste like dishwater if you are unlucky. And, while Veneto makes a lot of Pinot Grigio, it's usually better to get yours from Friuli-Venezia Giulia, where vines are grown on hills instead of flat fields. Amarone della Valpolicella is also from Veneto and it's a delicious wine made from grapes dried on straw mats after harvest, made primarily from a red grape called Corvina (Rondinella and Molinara are supporting players), then fermented into wine only after drying for two months.

A large swath of Prosecco also comes from Veneto, and it's worth noting that most of the Glera grape (used for Prosecco) is grown on flatlands. If you're seeking a high-quality bottling, get it from one of the two hilly DOCG regions instead: Conegliano Valdobbiadene or Asolo.

PIEDMONT

Located in the southern foothills of the Alps and the northern foothills of the Apennines (in Italian, *Piedmonte* means "foot of the mountain"), this region is hilly and dramatic, and the source of Italy's most famous red wines. Historically, the sunniest vineyard sites have been reserved for Nebbiolo, the star grape of Barolo and Barbaresco. Nebbiolo is markedly high in both acid and tannins, and it's not short on alcohol either—there is nothing like a young Barolo to make your mouth pucker! Other notable red grapes include Barbera (lower in tannin, fruitier than Nebbiolo, and bursting with energy when made by top producers—and a very good wine to pair with pizza, for the record!) and Dolcetto, which means "little sweet one" (but which I have honestly never found particularly sweet). When it comes to Piedmont's hillsides, soils are a mix of limestone and clay (which produce more delicate wines) and sandstone (which produce more structured and powerful ones).

On a separate front, Piedmont is also the region to thank for Moscato d'Asti, a sparkling, sweet, and perfumy wine that is not always taken seriously in the fine wine world but is fun, delightful, and tastes like a bowl of gardenias and lychee fruit. Moscato is also a force to be reckoned with. Of the more than twenty-seven million bottles produced annually, 80 percent goes to the US, with demand increasing, in part due to musicians from Drake to Lil' Kim referencing it in their songs.

LIGURIA

This region, located just south of Piedmont along the Mediterranean Sea in northeast Italy, is the home of places like Cinque Terre that pop up on your mood boards as well as a source of excellent wines, both white and red. Notable wineries like Punta Crena and Bruna have successfully preserved native grapes—delightful whites like Vermentino, Pigato, Mataòssu, and Lumassina. For me these always taste like the ocean, with plenty of citrus and mountain herbs, marked with a bit of salinity. The star red wine is Rossese di Dolceacqua, a light-bodied, bright, slightly spicy red made from Rossese grapes that grow on the steep slopes of the descending Apennines that jut into the Ligurian Sea. The Rossese grape is low in tannins, high in tart red fruit, and delightful even with dishes as delicate as crudo. Planted by the ancient Greeks, these wines have staying power—in Liguria and in your own life once you try them!

TUSCANY

Tuscany's identity has long been intertwined with viticulture. Located in central-western Italy, south of Bologna and north of Rome, Tuscany's rolling vine-covered hills span the outskirts of Florence to the Tyrrhenian Sea. Prized soil types throughout Tuscany are galestro, a calcareous marl that produces more delicate expressions of Sangiovese, and albarese, a sandstone that results in more structured styles.

Tuscany has a long history with viticulture; the delineated zone for Chianti, a hilly region in the center of Tuscany, dates back to 1716. Most of the region's viticultural identity involves Sangiovese, the thin-skinned, highly tannic, and acidic red grape that makes extraordinary wine in the right hands (Stella di Campalto, Poggio di Sotto, and Monteraponi all come to mind) and charming table wine the rest of the time.

The idea of "Super Tuscans," beginning in 1968 with the wine blend Sassicaia, grew from a notion that wines in Tuscany did not live up to their potential within the strict (but confounding) DOP laws—for example, that a percentage of white grapes, such as the not-always-amazing Trebbiano, had to be included in a red Chianti wine. A "Super Tuscan" refers to any wine made outside of the legal mandate of DOP regulations. While early Super Tuscans (Sassicaia, Tignanello, Ornellaia) celebrated Bordeaux grapes, often at the expense of regional varietals, later examples like Cepparello, from Paolo De Marchi's winery Isole e Olena, swung the other direction, celebrating Tuscany with 100 percent Sangiovese wine that was outside the law because it did not include the requisite white grapes.

MAKING SENSE OF MONTALCINO LABELS: AGING REQUIREMENTS

Rosso, Brunello, and Brunello Riserva are all made from the Sangiovese Grosso clone and can come from all of the same vineyards. The difference lies in the amount of aging each wine receives.

Wines labeled "**Rosso di Montalcino**" receive one year of aging (they cannot leave the winery before September 1 of the year after the harvest). These wines do not have to spend any time in oak. Subtext: This is a way for the winery to release an easy-drinking wine that's more approachable in youth and also generates some immediate cash flow.

Wines labeled "**Brunello di Montalcino**" must spend at least two years in oak barrels, plus an additional four months in bottle. The bottles may not be released before January 1 of the *fifth year* following the harvest. The mandatory bottle aging allows the wine to settle in and rest after "bottle shock" before it undergoes the strain of transport throughout the world.

Wines labeled "**Brunello Riserva**" must spend a minimum of two years in wood, plus six months in bottle, and they may not be sold before January 1 of the *sixth year* following the harvest. Riserva means that the winery invested in an additional year of aging the wine for the consumer.

Chianti Classico is the filet mignon of the Chianti zone and includes the four villages original to the Chianti zone before it was expanded (Radda, Greve, Gaiole, and Castellina). Just south of Chianti is Montalcino, a town atop a square hill. This slightly warmer area has also focused on Sangiovese in a particularly powerful clone, known as Sangiovese Grosso (or Brunello).

CAMPANIA

Located in the western ankle of Italy's boot, this region has an estimable winemaking track record dating back to 121 BCE. Today, it is best known for crisp, saline whites from the Falanghina grape, slightly waxier whites from the grape Fiano, and bold reds from Aglianico, a grape typically associated with the Taurasi appellation. Old Mastroberardino bottlings are treasures if you can find them, although the estate changed hands during a family feud in 1994 (so anything prior to that is superior!). Campania is also home to the Amalfi Coast—famous for its spectacular beauty dating all the way back to Homer's *Odyssey*, but also notable for making some life-changingly good wines thanks to producers like Marisa Cuomo, whose vineyards thrive in cliffs above the sea and whose top bottling, *Fiorduva*, is made from a mystery field blend of grapes from vines more than one hundred years old. It's breathtakingly good.

SICILY

Technically the largest island in the Mediterranean Ocean as well as one of Italy's twenty regions, Sicily has a millennia-old relationship with viticulture and remains a place of incredible quality and opportunity for discovery. A current wave of attention is incoming from winemakers settling on the slopes of Mount Etna, Europe's tallest active volcano, on the eastern part of the island. Sicily's climate is terrific for viticulture, with cool air from the ocean, volcanic soil with plenty of limestone pockets, and wind from the hills (which cover the island), making for balanced wines with a mineral backbone. Organic viticulture here has a four-thousand-year history, due to the sun and constant breezes.

Sicily also makes a beautiful and delightfully affordable range of wines beyond Etna, from red wines like Frappato, which grows in the southeastern part of the island on a bedrock of limestone and behaves almost like Pinot Noir. (Arianna Occhipinti put Frappato on the map in the early 2000s, and her uncle's winery, COS, makes beautiful examples as well.) And on Sicily's western coast near Marsala, Marco de Bartoli produces gorgeous wines from grapes traditionally used for marsala, like Zibibbo, a relative of the Muscat of Alexandria grape—not to be confused with perfumy Moscato d'Asti!

SPAIN

Spain has a very long history of winemaking, dating back to the Phoenicians in 1100 BCE, with evidence of vines dating back two millennia earlier, to 3000 BCE. The country has the most land allocated to vineyards (964,000 hectares, or 13.2 percent of the world's vines, ahead of France in second place and China in third place, respectively, as of 2021) and, while it experienced a somewhat recent slump in the mid-twentieth century under dictator Francisco Franco, Spain is once again making incredible wines with fresh perspective. As of 2021, Spain also has the most certified-organic vineyards by surface area—yet another reason to get excited about its wines!

While many regions—like Jerez, home to sherry—still carry on long-held winemaking traditions, the country endured some slumps brought on by one and a half centuries of Moorish rule (when alcohol was strictly forbidden), phylloxera, civil wars and world wars, and autocratic rule. That said, those periods also gave rise to some of the most exciting wines in the world today. Ever since 1975, when dictator Franco died and Spain became a democracy (and, in 1986, joined the European Union), the country has produced extremely compelling bottles.

Geographically and climatically, Spain is quite varied. From Galicia's green hills north of Portugal to the high-elevation Meseta Central (a plateau in central Spain averaging more than two thousand feet high) to a series of mountain ranges throughout the country, Spain benefits from plenty of sun and elevation, leading to those beautiful diurnal shifts that bring brightness and natural acidity to the wines.

Spanish food and wines are categorized by the DO (Denominación de Origen) system, similar to the AOP system in France and the DOP system in Italy. Spain's highest-quality wines are those labeled with a DO, with a special carve-out for the two top-tier regions, La Rioja and Priorat, classified as DOCa (or the DOQ—the same tier with a different acronym to account for Catalan spelling). While most DOs encompass regions, there are notable exceptions like Cava, a DO that spans the entire country of Spain despite most production occurring in Catalonia. (More on that on page 173.) My firm belief is that producers matter more than classification systems, in Spain or anywhere. The list below presents quality as defined by the Spanish government, but in the sections that follow, I share thoughts on producers worth seeking out as you explore Spain's wines.

A rough breakdown of wine Spain's classification system is as follows:

DOCA OR DOQ: Denominación de Origen Calificada is the highest level of Spanish wine classification for a region. *Calificada* means "qualified" and implies the highest-quality wine. Only two regions currently have this status: La Rioja and

Priorat, which use DOQ, short for Denominació d'Origen Qualificada, the more appropriate abbreviation in Catalan language.

DO: Denominación de Origen indicates the geographical origin and style of a wine. Wines labeled with a DO must conform to certain standards in both the vineyard and the winery, regulated by the region's governing body, the Consejo Regulador. Rías Baixas, Ribera del Duero, and Jerez are all examples of DOs. (There are about seventy in total.)

VP: Vino de Pago is a single-estate classification for high-end wineries that are basically overperformers in regions that haven't earned a DO.

VC: Vino de Calidad con Indicación Geográfica translates to "wine of quality with a geographical indication." This distinction is a stepping-stone between Vino de la Tierra and DO wines.

VT: Vino de la Tierra translates to "wine of the land" and is the lowest on the ladder—table wine.

One characteristic that has historically defined Spanish wines is an emphasis on aging for white, rosé (rosado), and red wines.

GALICIA

Galicia (*ga-LEE-thee-uh*), north of Portugal, is essentially an extension of that country's climate, with vineyards growing along the Atlantic coast. Often called Green Spain, the landscape is notably more vibrant than the oranges and browns present in much of Iberia. Galicia includes a handful of subregions, namely Rías Baixas, Ribera, Ribeira Sacra, Valdeorras, and Monterrei. Rías Baixas is the best known, famous for compelling whites made primarily from the Albariño grape that taste like a magical mix of tangerines, green apples, orange blossoms, river rocks, and ocean mist.

Ribeira Sacra, a relatively young region officially recognized only since 1996, is home to a dramatic landscape of steep, terraced slopes. Best known for its white wines made of Albariño and Godello (the latter of which is neutral and slightly savory, not dissimilar to Chardonnay), the region is becoming increasingly well known for its red wines made from the Mencía grape.

I was first introduced to mind-blowingly good wines from Galicia by way of the import portfolio of José Pastor Selections, which remains a wonderful resource for discovering some of my favorite Spanish producers. Listed alphabetically, Envínate, Guímaro, Nanclares y Prieto Viticultores, and Vimbio are all wine producers worthy of tracking down. I used to wait for allocations to post on Vervewine.com and snap them all up (sorry!). Both white and red wines from

Rioja wines labeled "Crianza," "Reserva," and "Gran Reserva" are all made from grapes within La Rioja in Spain and can refer to white, rosado (rosé), and red wines.

Crianza Tinto: minimum two years aging, including at least one year in (usually American) oak barrels for red wine

Reserva Tinto: minimum three years aging, including at least one year in (usually American) oak barrel and six months in bottle

Gran Reserva Tinto: minimum five years of aging with at least twenty-four months in (usually American) oak and at least twenty-four months in bottle

Galicia are terrific young, and they age well, too. I feel as though I've been beamed to sunny, green vineyards overlooking the Atlantic Ocean when I uncork these wines (especially the whites). They inspire me to build a meal around anchovies and pan con tomate.

LA RIOJA

Named for the river Oja, a tributary of the Ebro River, La Rioja has vineyards that predate the Romans, who built upon the preexisting infrastructure when they arrived in Spain. However, La Rioja's golden era did not arrive until phylloxera hit France, sending Bordeaux's winemakers and resources south, across the Pyrenees, to La Rioja, where a new railroad stop had just been installed. Previously landlocked and difficult to access, La Rioja benefited from the introduction of the railroad in 1880, providing the wines a route out of the region and over to France. La Rioja's most famous wineries—R. López de Heredia, CVNE, and La Rioja Alta—are all built next to the train station.

Geographically, La Rioja is divided into three areas: Rioja Alta (the westernmost subregion, known for both chalky, alluvial soil as well as iron-rich clay); Rioja Alavesa (the northernmost subregion and the smallest, with both the highest limestone content and the highest elevation), with vineyards planted almost entirely with Tempranillo; and lastly, the Rioja Oriental (previously Rioja Baja), the southeasternmost subregion and both the hottest and driest, with a more Mediterranean climate as opposed to the maritime influence of its more northerly neighbors. Garnacha has long been the main grape of the Rioja Oriental area and is commonly blended with Tempranillo from the other two subregions.

While La Rioja today is best known for its red wines, the region also produces complex, age-worthy whites (mainly from the Viura grape) and rosés (rosados) that are typically aged before they're released and are complex, waxy (in a good way!), and unlike wines from anywhere else. Aging requirements in La Rioja are longer than in other regions.

171

"OLD WORLD" COUNTRIES AND REGIONS

As of 2017, village names can now appear on bottles—and La Rioja's governing body, the Consejo Regulador, has created a new Viñedo Singular (single vineyard) category as well, made from hand-harvested grapes coming from single vineyard sites and adhering to lower yields and from vines older than thirty-five years.

Also new in 2017 is a category for sparkling wines, called Vino Espumoso de Calidad de Rioja, which must be made in the traditional method (see page 51).

CASTILLA Y LEÓN

Named for its past identity as the heart of an old kingdom, this large region located in north-central Spain takes up about one-fifth of the entire country. Castilla y León includes several subregions with ancient histories, some of which were codified as wine-growing areas as recently as 2007! The region is essentially a giant high-elevation plateau (2,600 to 3,300 feet) surrounded by mountains. Summers are hot and short, followed by long, cold winters where temperatures fall as low as -10 degrees Celsius (14 degrees Fahrenheit). Most regions follow the Duero River (the Spanish name for, and an extension of, the Douro River in Portugal), which helps moderate harsh temperatures throughout the year. Tempranillo, which goes by various aliases in the region—Tinta del País, Tinto Fino, Tinta de Toro, and Aragonês—is the star grape.

Bierzo, located closer to Galicia than the rest of Castilla y León, has developed a reputation for quality reds made from the Mencía grape grown on granite and slate soil. In 2017, the region introduced its own quality pyramid inspired by Burgundy's, with an emphasis on single-vineyard wines. Notably, the winemaker Álvaro Palacios, one of the advocates for Priorat who is partly responsible for its fame, has partnered with his nephew Ricardo Perez to shine a light on Bierzo.

Rueda, located south of the Duero River and farther east (inland) has recovered from post-phylloxera decline, thanks to the rediscovery of a native white grape, Verdejo, which thrives on the limestone soil here. Quite a bit of Sauvignon Blanc is planted here, too.

Ribera del Duero is the most famous subregion in Castilla y León, with a focus on Tempranillo (which covers 95 percent of vineyard land) and a viticultural history likely predating the Romans. Its history of exceptionalism can be traced back to one estate, Vega Sicilia, founded in 1864 (with a focus on Tempranillo, as well as the French varieties Cabernet Sauvignon, Merlot, and Malbec). Ribera del Duero suffered during the Spanish Civil War (1936–1939) and under Franco's dictatorial reign (1939–1975) but is now undergoing a renaissance and a renewed focus on quality, bolstered by the addition (in 1995) of winemaker Peter Sisseck's Pingus estate.

CATALONIA

Located along Spain's eastern coast and the Mediterranean Sea, Catalonia is a bit of a black sheep within the country. During Moorish rule, it was captured by Charlemagne and benefited from the king's interest in wine—he basically turned the region into an extension of Roussillon, in southwestern France. Barcelona is the region's capital, and the area has a delicious reputation for both wines and cuisine.

Catalonia's most significant wine contribution is the sparkling wine Cava, and 95 percent of the world's supply is produced here, though it can legally be made in seven of Spain's regions. Sparkling wine in Spain dates back to 1850, though Cava didn't officially get its name until 1970, named for the caves in which the wines are stored. (Previously it had been called Champaña, to the ire of the French, who took action that inspired the name change.)

All Cava is made in the traditional method (see page 51), and the three main classic grapes are Macabeu (Macabeo), Xarel·lo, and Parellada (all white grapes). Rosado versions of Cava also exist and include a minimum of 25 percent red grapes (mainly Garnacha Tinta, Pinot Noir, and Monastrell). A lot of Cava prioritizes quantity over quality, but there are new categories being introduced aimed at quality control.

I'm excited about the Cava DO's recent announcement that, as of 2025, all bottles labeled Cava de Guarda Superior must be made from organically grown grapes. Wines labeled Cava de Paratge Qualificat indicate that grapes were hand-harvested from a single vineyard.

Cava Aging Requirements

CAVA / CAVA DE GUARDA: minimum nine months on the lees from the day of tirage

CAVA DE GUARDA SUPERIOR: minimum eighteen months on the lees from the day of tirage; and made from organically farmed grapes

CAVA RESERVA: eighteen months on the lees from the day of tirage

GRAN RESERVA: minimum thirty months on the lees from the day of tirage (Brut, Extra Brut, and Brut Nature only)

CAVA DE PARAJE CALIFICADO: also known as the Cava de Paratge Qualificat in Catalan; minimum thirty-six months on the lees from the day of tirage; and hand-harvested from a single vineyard

PRIORAT

Catalonia's other star DO is Priorat, one of two DOCa regions (DOQ in Catalan), along with La Rioja, to hold Spain's top-tier classification. Named for the priory (or monastery) formed there in the twelfth century, Priorat gained its winemaking roots from a group of Carthusian monks who left Provence for Spain after hearing of a local shepherd's vision of angels ascending on a ladder to heaven. The monks set up shop at Scala Dei, or "Stairway to God."

Priorat's wine culture thrived during the Middle Ages before shrinking to a mere 500 hectares (about 1,235 acres) after phylloxera. The region's vineyard sites are steep and hard to maintain, things that both elevate the quality of wine potential there and also make the vineyards much harder to cultivate. Beginning in 1989, a group of winemakers led by René Barbier (of the winery Clos Mogador) committed to preserving Priorat's vineyards and, by 2000, the Catalan government upgraded the area to DOQ. By 2009, Spain's government recognized the region with its highest classification, the DOCa.

Priorat's soil, called llicorella, is a nutrient-poor, reddish-black slate—essentially fossilized volcanic ash compressed over time, with particles of mica that reflect sunlight and heat back onto the vines. Yields are very low, resulting in highly concentrated vines. Garnacha is particularly well suited to the area, as is Cariñena (Carignan). Tempranillo, Syrah, and Bordeaux varieties are also allowed, but the latter are frequently being replanted with Garnacha.

Priorat aging requirements are a minimum of twelve months in oak for Vi de Guarda classification, while all other classifications follow Spanish national requirements.

Additional Priorat Classifications

VI DE FINCA: single vineyard, independent of Vi de Vinya; 100 percent sourced from a single designated parcel

VI DE PARATGE: 100 percent sourced from a single zone

VI DE VINYA: quality vineyard designation given to single-parcel wines by the Consell Regulador, the governing body of Catalonia

VI DE GRAN VINYA: top-quality vineyard designation given to single-parcel wines by the Consell Regulador

VELLES VINYES: vineyard designation given to vineyards planted prior to 1945 (seventy-five-year-old or older vines)

ANDALUCÍA

This region, on Spain's southern coast, is one of the country's oldest and most historic, with Phoenicians (and along with them, grapevines) arriving in modern-day Cádiz around 1100 BCE. The region's focus is sherry and fortified wines. Andalucía is Spain's hottest region, with summer months averaging daily high temperatures near 100 degrees Fahrenheit. Three soil types are prominent in the area: albariza (bright white, chalky, porous, and limestone-rich—the most prized for Palomino grapes), barros (more fertile clay found in the valleys), and arenas (sandy soil found along the coasts where Moscatel, destined for sweet wine, is generally planted). Please see "Sherry," page 57, for more details.

CANARY ISLANDS

Closer to West Africa than to Europe, the Canary Islands are a series of volcanic isles making some of Spain's most exciting wines today. Wine-growing here dates back to the fifteenth century, when grapes were grown primarily to make a sweet, fortified wine from late-harvest grapes called sack that the Brits loved. Today, dry red wines made from Listán Negro, a grape native to the Canaries, are sommelier favorites, both for their charming accessibility and strong sense of place as well as for their incredible history (often century- or centuries-old ungrafted *Vitis vinifera* vines)—all combining to make soulful, delicious wines for prices lower than thirty dollars per bottle.

Envínate, a winery founded by four friends who met in oenology school, began vinifying indigenous grape varieties grown on vineyard parcels with interesting terroir from various regions in Spain. The winery helped bring mainstream awareness to the Canary Islands. (Yes, they also make wine in Galicia—this group works with compelling vineyards throughout Spain.)

Taken together and individually, the islands have a unique terroir. Growing techniques on the islands have evolved over the centuries. On the island of Lanzarote, which has a landscape comparable to the moon—mounds and craters of black volcanic topsoil—wineries have created hoyos (pits encased by stone semicircles forming small areas that block ferocious winds and allow vines to grow).

CHAPTER 7

"NEW WORLD" COUNTRIES AND REGIONS

I lived and worked in New York for seventeen formative years and became one of those people who got to eat, drink, and learn in one of the greatest food and wine cities in the world. Do I think New York is perfect? No, of course not. It has smelly subways and small apartments and giant cockroaches that still haunt me in my dreams. But it also has an extraordinary dining scene. All great wineries from all over the world want to have their wines in New York, and that is as true today as it was when I first began working in the city. New York is a phenomenal place to learn about wine because it has extraordinary access.

I bring this up because, after spending most of my paycheck for roughly a decade buying "classic examples" of wines around the world for exams and for pleasure, a majority of the great reference-point wines come from the "Old World."

In writing this chapter, I got very curious about how "New World" and "Old World" are defined and regarded. My own experience is inherently subjective, so I turned to *The Oxford Companion to Wine*, fifth edition, Jancis Robinson's (and now also Julia Harding's and Tara Q. Thomas's) tome that has a measured and thoughtful answer for pretty much any question about wine you've ever imagined. This quote from *The Oxford Companion to Wine* resonated, so I'm sharing it here:

> In the Old World, with its centuries of winemaking tradition, Nature is generally regarded as the determining, guiding force. In much of the New World, however, it was for long regarded with suspicion, as an enemy to be subdued, controlled, and mastered in all its detail, thanks to the insights provided by Science.

The *Oxford Companion to Wine* also explores "New World" countries as those who began to cultivate vines as a result of European colonization efforts. To this end, "Old World" countries are those that had established viticulture by the fourteenth century, while "New World" countries didn't get into their groove until the fifteenth century or much (much) later.

History has always fascinated me, and in the pages that follow I provide a snapshot of relevant events that helped shape each listed country's wine scene today. As is usual with wine, so many of the formative events involved wars and religion. Colonists needed wine for religious reasons, and while it's easy to think of the "New World's" wine scene as blossoming as recently as the mid-to-late twentieth century, all of today's major vine-growing areas had vineyards and a burgeoning wine culture by the 1850s. Mexico, Chile, and Peru all had vineyards by the mid-sixteenth century. South Africa's wine scene began in the mid-seventeenth century, and British settlers in the United States attempted to plant vines in Virginia in 1619, although their efforts failed, thanks to phylloxera and various diseases that *Vitis vinifera* vines couldn't withstand. On North America's opposite coast, in Baja California, none of the East Coast's vine diseases prohibited viticulture, and Spanish-Mexican-Jesuit missionaries established vineyards in the late 1670s. The British captain Arthur Phillip brought vines to Australia in 1788, and Samuel Marsden, an Australian missionary, imported them to New Zealand a couple of decades after that.

The countries below represent "New World" nations that I believe are making the most noteworthy and prevalent wines. I've listed the countries alphabetically. There are invariably some that I've left out or regions I'm not spending enough time on and wineries making great wines that haven't made it into these pages. The goal of this chapter is to provide some color and historical context for wines and regions most meaningful today. And if you know of more wineries doing great things, by all means let me know. I am very much still a student of wine, too.

ARGENTINA

Argentina is one of South America's most important wine-producing countries, along with Chile (the latter surpassed the former in production volume in 2023). Latin America's history with viticulture dates back to the sixteenth century, with the Spanish conquistadors' arrival in Mexico in 1521 and the Portuguese settlers' arrival in Brazil a decade later. By 1560, vines had spread to Peru, Chile, and Argentina.

Viticulture flourished, particularly in the foothills of the Andes Mountains. Spanish conquistador Pedro del Castillo founded Mendoza in 1561, and while King Felipe II of Spain issued a law banning wine production in South America in 1595, intending to redirect support back to Iberian producers, the law had a special carve-out for vines grown in connection with the Catholic Church. As a result, the law had an unintended effect of jump-starting innovation and quality in South America. Its wine industry prospered like never before for almost three hundred years, thanks to the monasteries' power and resources. The law dissolved in the eighteenth century, along with Spanish political control—and the nineteenth century brought waves of European immigrants, a new influx of viticultural knowledge, and the railroad, which helped rapidly expand the continent's wine scene (particularly in Argentina, where it connected Buenos Aires to Mendoza in 1885).

Argentina has a long history of wine consumption per capita, with about 75 percent of all production staying inside the country. Exportation wasn't significant until a series of political and economic calamities befell the country during the twentieth century, which finally began stabilizing in the 1980s before a steep recession in the late 1990s. (This devalued the Argentinian peso by 300 percent, but on a positive note, the recession attracted foreign investment and a new export market.)

Argentina is divided into three winemaking provinces: the highest-elevation northwest provinces, Cuyo's central provinces (where Mendoza is located and more than 75 percent of all Argentinian wine is produced), and the southern province of Patagonia, which includes Río Negro and Neuquén. In terms of wine law and label requirements, Argentina's governing body for all things wine and viticulture is called the Instituto Nacional de Vitivinicultura (INV). Since 1999, Argentina recognizes three levels of quality, all geographically oriented, like the United States' AVA system (which we will cover later in this chapter). The INV's three levels of quality, in ascending order, are below:

1. **IP (INDICACIÓN DE PROCEDENCIA):** table wines containing at least 80 percent of grapes from the IP region

2. **IG (INDICACIÓN GEOGRÁFICA):** higher-quality wines grown, vinified, and bottled in a designated area

3. **DOC (DENOMINACIÓN DE ORIGEN CONTROLADA:** according to Wines of Argentina's website, DOCs are IGs with wine style relegations, such as yield and aging requirements, as in Europe. There are only two DOCs, and both are located in Mendoza: Luján de Cuyo (for Malbec only) and San Rafael.

Argentina occupies a large southeastern portion of the South American continent and is home to many of the world's highest-elevation vineyards, with some vines growing at 3,000 meters (nearly 10,000 feet) above sea level. Most vineyards run along the foothills of the Andes. They have access to water, thanks to irrigation channels built by the now-extinct indigenous Huarpe population, who created them for agricultural purposes prior to their deportation to Chile in the eighteenth century. The irrigation channels capture snowmelt from the mountains and direct the water to vineyards, making viticulture possible.

Climatically, most of the country is considered continental. Vineyards benefit from a rain shadow effect, and temperatures are generally hot and dry. Vineyards also profit from a warm, powerful wind called the Zonda, which blows down from the mountains and eliminates mildew by keeping vines dry. One side effect of the Zonda is that grape skins grow thicker in order to protect the fruit, often resulting in higher tannins and more structure than wine made from those same grape varieties grown elsewhere. However, thanks to the Zonda, organic viticulture is easier as the wind blows away humidity, keeping downy and powdery mildew at bay. Hail is a threat in both spring and summer. Wineries commonly blend grapes from various vineyards throughout a region to mitigate against the risk of hail wiping out an entire plot. Most vines are grown on soil with a high amount of sand, which has allowed Argentina to maintain many vineyards of ungrafted vines.

NORTHWESTERN PROVINCES:
SALTA, JUJUY, CATAMARCA, AND TUCUMÁN

These northern regions have some of the highest elevations in the world. While increasingly refined wines are being made across the four provinces, Salta is the most famous. Donald Hess's vineyard, Colomé Altura Máxima, is planted at 10,206 feet elevation. (I've never tasted the wine, so this isn't an endorsement—more of a curiosity that vines can actually grow that high.) The high elevation and northerly location provide a compelling mix of sunlight, ultraviolet light, and plenty of cold winds blowing from the Andes, leading to wines that are very ripe and have good natural acidity.

SAN JUAN AND LA RIOJA

The San Juan region borders Mendoza to the north, and La Rioja (not to be confused with La Rioja in Spain) is Argentina's oldest winemaking region, located north and northeast of San Juan. While La Rioja's historic significance is noteworthy, today most production is controlled by a quantity-oriented co-op and is not well known outside of Argentina.

MENDOZA

This province, located in central western Argentina just south of San Juan, produces three-quarters of the country's wine and largely focuses on Malbec, which has captured international attention with its dark fruit characteristics, moderate tannins, and bold flavors.

Vineyard elevations average about 2,400 feet. The Mendoza province is divided into north, central, south, and east sectors, as well as the Uco Valley to the west, home to the province's highest vines. Soils are a mix of alluvial sand and clay, and irrigation is essential. The Uco Valley is a rising star in Argentina, admired for its cool temperatures and poor soil (resulting in more interesting wine) composed of large, calcium-rich stones along with sand and loam. The Uco Valley, for context, is roughly the size of Burgundy. Other more established parts of Mendoza include Luján de Cuyo (Argentina's first controlled appellation, which allows only Malbec) and the Maipú Department, which is warmer and lower in elevation. Some of the region's most historic and quality-oriented wineries include Bodega Catena Zapata, Susana Balbo, and Terrazas de los Andes (the latter is owned by LVMH).

PATAGONIA: RÍO NEGRO AND NEUQUÉN

These provinces are less famous, but they're responsible for some exceptional Argentinian wines. A lot of Malbec can strike me as cookie-cutter, whereas wines from Río Negro and Neuquén taste a bit wilder, with plenty of soul. These southern provinces in Patagonia are significantly cooler and slightly more humid than the better-known regions in the north, although Río Negro in particular has a long history as Argentina's best region for fruit production, and today it is responsible for some of Argentina's most exciting wines. Countess Noemi Marone Cinzano and Piero Incisa della Rocchetta (of Sassicaia family fame) both visited the region and immediately saw its potential. Bodega Noemiá, founded by the countess, makes exceptional Malbec, and Piero Incisa's estate, Bodega Chacra, prioritizes own-rooted (ungrafted) old-vine Pinot Noir, farmed biodynamically. I was able to work a harvest at Bodega Chacra a few years ago, which only furthered my respect for the land and the wines.

AUSTRALIA

Australia's viticultural history dates back to 1788, with Captain Arthur Phillip and his fleet of British prisoners. They planted vines they'd picked up in Cape Town en route to a small penal colony, which today is the city of Sydney. Vines spread to Tasmania, South Australia, Victoria, and Western Australia by 1830, at which point European immigrants began to arrive and viticulture started improving. During the 1850s, Victoria became Australia's leading area for wine production, largely driven by an influx of settlers searching for gold. This lasted until the late 1870s, with the discovery of phylloxera in the region. Phylloxera devastated vines and halted production in Victoria but paved the way for a flourishing wine scene in other parts of the country, notably South Australia. The epicenter of wine production shifted from Victoria to Barossa and to the wide, hot, and newly irrigated areas north of it, which mostly churned out fortified wine. For much of the twentieth century (until 1960), 80 percent of Australia's exports consisted of sweet, fortified wines. If that number seems ridiculously large, consider that the US, a critical export market for Australia (and number one by export value still today) was at that time still drinking Thunderbird—Gallo's concoction of lemon juice concentrate and white port—more than any other wine-based beverage.

Preferences for dry wines emerged in the 1970s, which coincided with new technologies that mechanized both winemaking and vine growing, an important shift in remote Australia, where labor was (and remains) a constant challenge. Australia's mechanized approach allowed it to produce large quantities of soft, fruity wines—especially Shiraz (Syrah) and Chardonnay—in the early 2000s. These became instant hits in both the US and the UK. "Critter wines," which are mass-produced, inexpensive wines like [yellow tail], had their fifteen minutes of fame before the "millennium drought" from 2001 to 2009 rocked Australia, exacerbated by climate change. All of this called into question the longevity of cheap Australian wines that required massive amounts of irrigation.

At the opposite end of the quality spectrum, Australia profited from technological advances and global attention during the 1990s—as well as from the trend toward "powerful" wines promoted by American wine critic Robert Parker and his 100-point rating system. The Australian auction house Langtons began publishing a guide in 1990 that listed the country's "Exceptional" wines meant for long-term collecting. It included Penfolds Grange, a Shiraz-driven blend of grapes from multiple vineyards first produced in 1951, plus single-vineyard bottlings (more and more the norm for excellence in Australia), including Henschke's "Hill of Grace" Shiraz, Grosset's "Polish Hill" Riesling, and others, with a focus on juicy, powerful reds.

Today, a shift in sommelier and consumer preferences from big, juicy red wines with high critic scores to more elegant, artisan-driven wines has created challenges for Australia's export industry. That said, the shift opens the door for more focus on some of Australia's cooler regions, like Western Australia and Tasmania, and offers a chance for smaller estates more in line with today's consumer interests to thrive.

Notably, mechanization and an emphasis on both science and cellar hygiene, or an obsession with "clean" wines, have been criticized by critics and sommeliers alike for robbing wines of a sense of place—but these contributions have also benefited the global wine community. We can thank Australia (specifically the Australian Wine Research Institute in Adelaide) for decoding the *Brettanomyces* genome (a fault—or attribute—long associated with a barnyard aroma in wine) in 2011, as well as for identifying the peppery smell in Shiraz (Syrah) as rotundone (see page 34). More broadly, Australia's research has given names and scientific backing to what had previously been considered part of a wine's mystery and romance.

Below is an overview of Australia's most important states for wine production today.

AUSTRALIAN WINE LAW

With a geographical landscape so vast and varied, Australia is able to produce every style of wine—from still and sparkling to fortified and sweet. Wines are labeled with designations called Geographical Indications, or GIs, which are purely geographical and do not mandate any parameters around grape varieties, yields, or practices within the vineyard or winery. Many of Australia's historic label terms—such as Tokay (for sweet wines) and Hermitage (for Shiraz)—were taken directly from European regions, a source of consternation finally eliminated when the EU mandated their dissolution, with the last terms phased off of Australian labels in 2020.

AUSTRALIA
WINE REGIONS
PACIFIC OCEAN
GREAT BARRIER REEF
NORTHERN TERRITORY
GREAT SANDY DESERT
WESTERN AUSTRALIA
QUEENSLAND
GREAT VICTORIA DESERT
SOUTH AUSTRALIA
NEW SOUTH WALES
Clare Valley
Barossa Valley
Eden Valley
Adelaide Hills
McLaren Vale
Murray Darling
Hunter
Sydney
Riverina
Swan Hill
Perth
Margaret River
VICTORIA
Coonawarra
Grampians
Melbourne
Mornington Peninsula
Macedon Ranges
Geelong
Yarra Valley
Hobart
INDIAN OCEAN

WESTERN AUSTRALIA

This state is Australia's farthest west, as the name suggests, with its original sub-regions bordering Perth, the capital city. The most famous subregion, Margaret River, however, is a new addition dating back just to 1965, when a local agronomist assessed the land and recommended viticulture. Margaret River is surrounded by water on three sides, a peninsula jutting out into where the Indian Ocean and Southern Ocean meet. In addition to wine, the area is known for spectacular beaches, inland rivers, and ancient underground caves with fossilized mammals dating back forty-six thousand years.

The "first five" wineries, as they are nicknamed, comprise some of Australia's most notable wineries today. These include Vasse Felix, Cullen, Leeuwin Estate, and Cape Mentelle, along with Voyager and Moss Wood, which end up as the fifth depending on whom you're consulting. In addition to its Chardonnay, which is world-class, Margaret River is becoming known as one of Australia's best sources for Cabernet Sauvignon. According to locals, the sources of Margaret River Cabernet can be traced to the state's first plantings, from an estate called Swan Valley, which came directly from the original cuttings that Captain Arthur Phillip picked up in Cape Town on his way to Australia back in 1788. The red gravelly loam soil in Western Australia is a terrific match for the grape, and the crisp ocean breeze ensures that wines are bright and balanced.

SOUTH AUSTRALIA

The continent's most famous and prolific state, South Australia currently produces 50 percent of Australia's wine (down from 75 percent in the 1940s). Its reputation rose in the late 1800s, when it implemented strict rules against importing vines after phylloxera arrived in Victoria. South Australia includes its capital city, Adelaide, home to the University of Adelaide, Australia's celebrated viticulture and oenology university. The state's most prominent subregions include the Barossa zone, Fleurieu, and the Mount Lofty Ranges. The Barossa zone, named for a Spanish battlefield during the Napoleonic Wars, includes both the Barossa GI and Eden Valley GI, home to some of Australia's most respected vineyards. Temperatures are warm, the valley floor is flat, and soils are loamy clay, with underground water reserves crucial for irrigation.

Shiraz is the star in Barossa Valley, accounting for half of all vines planted. The regional style is juicy, bold, concentrated, and high in alcohol, rarely below 15 percent. Because South Australia remains phylloxera-free, some of the world's oldest Shiraz (and other) vines are located here. The four terms below can be found on labels, corresponding to vine age:

OLD: at least thirty-five years old

SURVIVOR: at least seventy years old

CENTENARIAN: at least one hundred years old

ANCESTOR: at least one hundred twenty-five years old (how cool is that?!)

Dry Shiraz is the most recognized example, but a regional favorite (exported to the US to a small degree) is sparkling Shiraz, usually off-dry and not a terrible idea for a dessert pairing if figs and chocolate are on the menu.

East of and contiguous with Barossa Valley lies Eden Valley, which has a higher elevation, cooler temperatures, and more granite in the soil. The cooler areas have become known for very good, zippy Rieslings in the south and for elegant, spicy expressions of Shiraz in the north. Australia's most famous red wine, Henschke's "Hill of Grace" Shiraz, is grown here, just across the border from Barossa. Just north of Barossa lies the temperate Clare Valley, known for terrific dry Rieslings, plus some Shiraz and Cabernet. The McLaren Vale GI is located in this state, too, and is known for powerful red wines made from Cabernet Sauvignon. Coonawarra, located on the Limestone Coast, is renowned for its Cabernet Sauvignon grown on "terra rossa"—red clay loam on a bedrock of limestone.

VICTORIA

The smallest and coolest state on mainland Australia, Victoria has one of the longest histories with wine, dominating production until phylloxera crippled the industry there in the 1870s. (The Victorian government required every infected vine pulled up and destroyed.) Pinot Noir, Chardonnay, and Syrah/Shiraz all thrive here, thanks to cool Antarctic ocean wind currents that temper long, sunny days and keep acidity sharp in the wines.

The Port Phillip zone—which lines the coast of the Indian Ocean around the state's capital city, Melbourne, and encompasses the regions of Geelong, Macedon Ranges, Mornington Peninsula, Sunbury, and Yarra Valley—represents both historic Victoria as well as the state's most noteworthy wines today. Farther north, temperatures are much warmer, and viticulture is possible thanks to irrigation. Tasting Pinot Noir from the winery By Farr, grown on a limestone and volcanic soil patch in Geelong's Moorabool Valley, was an aha moment for me—I first tried the wine blind at a dinner during the harvest at Domaine Dujac and incorrectly identified it as Domaine de la Romanée-Conti (I was not alone in that guess!). As it happens, most of the Pinot Noir clones in the region were brought over in 1831 from Burgundy's famous Grand Cru vineyard, the Clos de Vougeot.

CHILE
WINE REGIONS
BOLIVIA
PARAGUAY
BRAZIL
URUGUAY
ARGENTINA
PACIFIC OCEAN
ATLANTIC OCEAN
Copiapo
Atacama
Coquimbo
Valparaiso
Aconcagua
Aconcagua
Santiago
Central Valley
Concepcion
Bio Bio
South Region
Austral Region

TASMANIA

Tasmania is an island off of Victoria's coastline and is Australia's southernmost and coolest region. Best known for Chardonnay, Riesling, and Pinot Noir, the climate is often compared to Germany's Rheingau and France's Champagne. Half of all bottles produced on the island are sparkling. Tasmania's best-known winery is Jansz, which specializes in sparkling wines made in the traditional style and was founded in the 1980s as a joint venture by Louis Roederer and Heemskerk.

CHILE

Chile has produced wine since the 1540s, after Spanish conquistador Francisco Pizarro and his lieutenants brought vines to Peru in the 1530s. Then they traveled south, seized land from the native Mapuche inhabitants, started wars, planted vines, and ultimately turned the locals into slaves. Soon after that, in the mid-sixteenth century, Catholic missionaries brought Moscatel (also known as Muscat of Alexandria) and Listán Prieto, a grape from the Canary Islands (now known in Chile as País and in California as Mission). Viticulture thrived, even during Spain's ban on "New World" wine production, thanks to the carve-out for wine produced for sacrament (and also because no one on the ground particularly cared to enforce the order).

The early nineteenth century marked Chile's independence, as well as its exposure to and growing admiration for French wines. Napoleon had installed his brother as the monarch of Chile in 1808, which inspired revolt and the subsequent Chilean Declaration of Independence a decade later. Fortuitously, a French natural-ist named Claudio Gay encouraged Chile's government to import samples of exotic European plants, including grapevines, before phylloxera and powdery mildew ravaged Europe and the rest of the wine-growing world. Once free of European rule, wealthy Chileans (who made fortunes in mining the country's rich mineral deposits) began traveling to France and picked up inspiration from Bordeaux's lavish "château culture." One man in particular, Don Silvestre Ochagavía Errázuriz, imported both French vines and a French winemaker, founding Viña Ochagavía, an estate that remains in operation today. He is still referred to as "the father of Chilean wine."

Chile's wine culture today is a duality, with large, wealthy, family-run, generations-old estates producing the lion's share of exports and a new, iconoclas-tic generation celebrating non-French grape varieties and taking a terroir-driven, minimal-intervention approach. Many of the most famous Chilean exports are large corporations, including partnerships with First Growth Bordeaux and Napa

Valley estates. Today, 80 percent of Chile's production is condensed within four wineries: Concho y Toro, Santa Rita, Santa Carolina, and San Pedro. But bigger doesn't mean better, and there is also a growing community of small independent winemakers looking to preserve centuries-old traditions before they are lost. MOVI, a quickly growing group founded in 2009, stands for Movimiento de Viñateros Independientes ("Movement of Independent Winemakers") and defines itself as "an association of small quality-oriented Chilean wineries who come together to share a common goal to make wine personally, on a human scale."

Bordered by the Pacific Ocean to the west and the Andes Mountains to the east, Chile's physical barriers, plus Claudio Gay's collection of pre-phylloxera *Vitis vinifera* vines, have made it the only major wine-producing nation not affected by phylloxera. (The bugs can't fly over the tall peaks of the Andes.) The country is long—just shy of 3,000 miles—and narrow (less than 60 miles wide at its thinnest point), and 80 percent of the country is covered with mountains. The Pacific Ocean cools the entire western coast, with a particularly cool influence coming from the Humboldt Current to the south. Snowmelt from the Andes ensures grapes have plenty of water—and irrigation is essential. Generally speaking, the long, sunny days combined with the cooling factor from the Pacific create bold, ripe fruit and balanced natural acidity.

As of 2012, the Chilean Ministry of Agriculture introduced three new geographic terms that can be added to existing regions and used on wine labels: Costa (the coast), Entre Cordilleras (between the hills), and Andes. Standout wines I've tasted from Chile include examples from Casa Marin in the San Antonio Valley, west of Maipo and along the Pacific coast, as well as from Louis-Antoine Luyt, founded by a Frenchman of the same name who spent time working with legendary Burgundy producers Philippe Pacalet and Marcel Lapierre and who now makes small quantities of exceptional wines from hundred-year-old País vines that are attracting enthusiastic international acclaim. Most notable, perhaps, is Pedro Parra, a winemaker and consultant with a PhD in terroir viticoles from the Institut

CHILEAN GRAPE VARIETIES

Demonstrating the impact of French influence, the two most prominent varieties grown in Chile today are Cabernet Sauvignon and Sauvignon Blanc, eclipsing the País and Moscatel originally planted by Spanish colonists. Merlot has a long history in Chile, too, though it tended to strike those who tasted it as "green," a descriptor that made more sense after modern DNA testing revealed that half of what was thought to have been Merlot in Chile was actually Carmenère, which has more pyrazenes and a greener, bell-peppery taste. Chardonnay is increasingly planted and shows a lot of promise—large Burgundy estates like William Fèvre, in Chablis, have started joint ventures here, as have other Burgundians.

National Agronomique Paris-Grignon. Parra has worked with top wineries across the globe and counts world-renowned vignerons Jean-Marc Roulot (Meursault), Louis-Michel Liger-Belair (Vosne-Romanée), Biondi-Santi (Montalcino), and Comando G (near Madrid) as clients, among others. Parra has established a winery, Jarilla Wines, in Itata, the southern part of his home country, Chile, where he cultivates organically farmed Cinsault and País.

ATACAMA

This far-north region is dry and desert-like, generally reserved for Pisco (a brandy made from fermented grapes) as opposed to exported varietal wines.

COQUIMBO

This region is arid and requires irrigation but is growing in popularity. Star subregions include the Elqui and Limarí valleys. Limarí has limestone soil and is increasingly planted with Chardonnay, while Elqui producers are investing heavily in Syrah.

ACONCAGUA

Aconcagua boasts alluvial soils and is increasingly planted with Bordeaux varietals. The prominent estate Errázuriz is based here and its top wine, "Seña," notably beat both Château Lafite Rothschild and Château Margaux in 2004 at a blind tasting in Berlin. In the Casablanca subregion, located close to the coast, the Humboldt Current keeps temperatures cool and adds a burst of acidity to the wines. Crisp Sauvignon Blancs and Chardonnays from here are becoming widely distributed.

VALLE CENTRAL (CENTRAL VALLEY)

This is Chile's most famous and established DO, with subregions like Maipo Valley, home to large, well-funded estates like Concho y Toro. Red grapes, led by Cabernet Sauvignon, are the most prominent.

NEW ZEALAND

One of the newest of "New World" regions, New Zealand's wine scene was strangled by regulation until essentially the late 1970s. The Māori people, believed to have arrived on the North Island from Tahiti—by canoe!—in the early fourteenth century, were the first humans to live on the islands. They were followed by the British, who did not bring viticultural expertise when they arrived in 1819. Scottish-born James Busby brought vines from Australia to New Zealand in 1833, and French settlers began to establish vineyards in Hawke's Bay in 1851. Momentum for a wine culture began to build, however slowly, when a group of Dalmatian people (from what is Croatia today) landed in the early twentieth century and planted vineyards in West Auckland. Despite the growing interest in viticulture, a strong temperance movement gained steam. Local New Zealanders campaigned to ban alcohol, and it was only soldiers voting abroad from Europe after World War I who tipped the balance in favor of continued legalization—although booze was hardly easily accessible. Until 1967, restaurants and bars were not allowed to sell alcohol after 6 p.m., and wineries themselves could not sell even a glass of wine until 1976. Grocery stores, which today account for 60 percent of all wine sold in the country, finally sold wine legally beginning in 1989. New Zealand's flourishing wine industry is essentially as old as Taylor Swift.

While New Zealand has maintained a reputation for quality, it produces very little quantity in the grand scheme—just 0.5 percent of the world's wine as of 2020. The country is the world's thirty-first-most productive, between Ukraine and Mexico. In 2016, New Zealand codified eighteen wine regions in its Geographical Indications Bill, which protects the use of regional names like Marlborough, Central Otago, and Auckland. Given the importance of the export market, New Zealand subscribes to the international 85-percent rule: if a grape, vintage, or geographic origin is mentioned on the label, at least 85 percent of the wine must come from these qualifiers.

NORTH ISLAND

While it was the first to grow vines, the North Island produces much less wine than the South Island and is far less suited to viticulture on account of its subtropical climate. Still, it is responsible for some excellent Bordeaux blends and Syrah, thanks to well-draining, gravelly soil called Omahu. The North Island's three most important growing regions are Hawke's Bay, Wairarapa, and Gimblett Gravels. A few richer styles of Chardonnay and Sauvignon Blanc are produced here, too.

NEW ZEALAND
WINE REGIONS

Northland

● **Auckland**

Waikato / Bay of Plenty

Gisbourne

NORTH ISLAND

Hawkes Bay

PACIFIC
OCEAN

Wellington
■ **Wellington**

Nelson
● **Nelson**

Marlborough

SOUTH ISLAND

● **Christchurch**

**Cantebury/
Waipara**

Queenstown ●

Central Otago

● **Dunedin**

PACIFIC
OCEAN

SOUTH AFRICA
WINE REGIONS

SOUTH
AFRICA

Cape
Town

WESTERN CAPE

Lutzville Valley

ATLANTIC
OCEAN

Olifants

Citrusdal
Mountain

Citrusdal
Valley

Sutherland Karoo

Swartland

Ceres
Plateau

Worcester

Tulbagh

Darling

Wellington Breedekloof

Langeberg Garcia

Cape Town

Robertson

Plettenberg Bay

Paarl

**Cape
Town**

Stellenbosch

Swellendam

FALSE
BAY

Elgin

Overberg

Breede

WALKER
BAY

Walker Bay

Cape Agulhas

BENGUELA
CURRENT

INDIAN OCEAN

Today, most of New Zealand's vines—a full 70 percent—come from the region of Marlborough, on the South Island—and 60 percent of grapes planted in the country are Sauvignon Blanc. The South Island is larger and better suited to viticulture than the North Island. This is largely thanks to the Southern Alps, a mountain range that runs north to south, creating a rain shadow effect to the east and allowing plenty of sunshine in what would otherwise be a cool, southerly region. Marlborough Sauvignon Blanc has captured the world's interest with its bright, citrusy, and gooseberry notes. Its success, however, is quite recent; the first vines were planted in 1974!

In addition to Marlborough Sauvignon Blanc, the South Island produces small amounts of very good Chardonnay in Nelson, to the west. The country's most important region for red wine, however, is Central Otago, the southernmost growing region on Earth. Central Otago is also New Zealand's highest in elevation. Pinot Noir is the star, benefiting from long days and plenty of sunlight for ripening, plus cool breezes from the Antarctic that add balance and zip. It is also the only truly continental climate in the country. Top producers include Felton Road, Rippon Vineyard, and Mt Difficulty, who also make Riesling and Chardonnay along with the trademark Pinot Noir.

SOUTH AFRICA

South Africa has a long and often complicated history with wine, beginning in 1652 with a thirty-something Dutch surgeon named Jan van Riebeeck, the country's first European settler. In 1652, vine cuttings taken from western France were planted in what is now Cape Town. Seven years later, in 1659, South Africa's first wines were pressed from those vines. South Africa's history with wine flourished through the eighteenth century, with highlights like Klein Constantia still globally renowned three hundred years later. But its history is also marred by a series of calamities involving overproduction, phylloxera, power struggles between the British and the Dutch, a powerful and misguided governing co-op, global boycotts for much of the twentieth century, and, more nefariously, injustices in the power structure of who was allowed to make wine or own vineyards (only white people, until apartheid ended in 1994). Once apartheid ended, South African growers and consumers had access to wines outside of their borders again, and a renaissance began. The country is now hitting its stride with some world-class examples. That said, the industry still struggles—bulk wine remains a major aspect of the country's production, with co-ops still holding power and only 15 percent of independent wineries reporting profitability.

South Africa is a strikingly beautiful country, with mountains jutting up out of the ocean and dramatic peaks capped with snow. It sits at the meeting point of the Atlantic and Indian oceans, two bodies of water that have profound impacts on the country's climate and temperature. South Africa is far enough from the equator to experience cool temperatures and a long growing season (grapes have time to mature, growing complex and delicious) thanks to the chilly Benguela Current, a powerful and cold upwelling current from the Antarctic that cools South Africa's southern coastlines, preventing rot and mildew in nearby growing regions.

The country has long prioritized white grape varieties. Prior to phylloxera, Sémillon was the most-planted grape, although today Chenin Blanc (locally called Steen) has taken that spot, accounting for 20 percent of all vineyard land. Sauvignon Blanc comes second, and in third place, Chardonnay.

One benchmark of South African excellence in wine is A.A. Badenhorst, led by Adi Badenhorst, who is an innovator and thought leader not just in South Africa but in the global sphere. He has worked at renowned wineries worldwide, from Wither Hills in New Zealand to Domaine Alain Graillot and Château Angelus in France, but his family roots link back to Klein Constantia. He began redefining quality in South Africa today while helping to restore its identity on the global stage.

Post-apartheid, South Africa's red wines are also gaining a reputation for excellence. The most-planted grapes are Cabernet Sauvignon and Syrah (both account for about 10 percent of vineyard land), with Pinotage—the country's contribution to grape varieties by crossing Pinot Noir and Cinsault (which, in South Africa, is called Hermitage)—on the decline. Pinotage never captured international imagination, probably because it unfortunately tastes like burnt tires.

Despite quality perception issues, South Africa has some of the strictest regulations in the "New World," beginning in 1973 and predating America's own AVAs (American Viticultural Areas), which didn't come about until the 1980s. South Africa's classification is called WO (Wines of Origin), and the system goes to great lengths to codify everything down to single vineyards (which must be inspected, approved, and not larger than six hectares—one hectare is approximately the size of two and a half football fields). While the WO system makes no mandate on vineyard or winery practices, it does require any wine stating a place of origin to be 100 percent from that place, and in order to list vintage or grape variety, a tasting panel has to approve it.

South Africa is divided into six geographical regions, with Western Cape being the only one relevant for viticulture. Within the Western Cape there are five production regions: Breede River Valley, Cape South Coast, the Coastal Region, Klein Karoo, and Olifants River. Only the two most southerly regions, the Coastal

Region and Cape South Coast, are known for quality wine. (The others are associated with co-ops and bulk production.) Within the regions are smaller areas called districts, and below that are smaller areas called wards (which sometimes exist outside of districts). The smallest unit areas are single vineyards.

Swartland and Stellenbosch (of the Coastal Region) and Walker Bay and Elgin (of Cape South Coast) all line the coast and have developed reputations for the highest-quality vineyard land. They are also, generally speaking, home to the most talented and ambitious winemakers. As of February 2020, an entirely new WO exists outside of previous categories—the Cape West Coast, producing wines characteristic of a more coastal influence.

SWARTLAND

While the region has a long history with wine, it's more removed and remote from Cape Town and was primarily co-op controlled throughout the twentieth century. One winery, Spice Route, sought to change that; its proprietor, Charles Beck, planted Mediterranean grape varieties (Grenache, Mourvèdre, Shiraz) and invested in viticulture in a way others had not done in the region for centuries. Spice Route's more notable claim to fame is having employed Eben Sadie, a man who is credited with turning Swartland into an epicenter of thoughtful wine-making that has captured global attention. Sadie left Spice Route and inspired some friends to join him in the Swartland. Today, they have founded the SIP, or Swartland Independent Producers, which has an established mission to "build a strong regional identity by focusing on what works best," with core values and requirements for certification, including approved grape varieties and a commitment to little to no intervention in the cellar. (In short, the SIP is bringing a very "Old World" philosophy in this very newly invigorated "New World" region, and the wines are terrific.)

STELLENBOSCH

Cape Town's eastern neighbor, Stellenbosch, is the heartbeat of South Africa's wine industry, with a history dating back to the seventeenth century and a location less than an hour's drive from the capital city. Benefiting from both a robust tourism industry and an oenology school, wineries in the region have unique access to distribution in Cape Town.

Simonsberg-Stellenbosch, the ward farthest inland and farthest north, focuses on red grapes, including Cabernet Sauvignon, Merlot, Pinotage, and Shiraz, which profit from the warmer temperatures and clay soil. The Bottelary ward, to the west, is closer to the coast and therefore cooler, specializing in Chenin Blanc and

Sauvignon Blanc in addition to the aforementioned red varieties. My friend and former harvest-mate Mick Craven now makes beautiful wines in Stellenbosch with his talented wife, Jeanine. I love the Craven Wines lineup and am happy to say I'd like it just as much even if I didn't know them and think of them as wonderful people!

WALKER BAY

This district in the Cape South Coast is home to three neighboring wards collectively referred to as Hemel-en-Aarde in Afrikaans ("Heaven on Earth"). Some new and quite skilled winemakers specializing primarily in Chardonnay and Pinot Noir are investing here, largely inspired by the Hamilton Russell Vineyards estate, founded sort of illegally in the 1970s. Hamilton Russell made wine so far superior to anything else being produced in the area that the powerful government-backed KWV co-op, which did not allow Pinot Noir in the area, had to ultimately relax its paralyzing grip on the country's wine industry before privatizing in the 1990s. (Prior to this, the KWV once forced Hamilton Russell to dump five thousand liters of wine down the drain, before court battles and public opinion finally changed the co-op's mind.) Elgin, another district in the Cape South Coast, is producing quality Sauvignon Blanc, Chardonnay, Merlot, and even some Riesling.

UNITED STATES

While the United States has been producing reliable quality wine for only a handful of decades, it's had an outsized impact on the international stage. Today, the United States is the world's fourth-largest producer, behind France, Italy, and Spain—and it is the largest consumer of wine worldwide, having surpassed France in 2011. From modern winemaking techniques encouraged by the University of California at Davis to critics like Robert Parker and his oft-replicated 100-point rating system, America changed, for better or worse, the way much of the world drinks. While the point system still looms large, the homogeneity it created is beginning to wane, thanks to a renewed celebration of individuality in viticulture and terroir, as well as a consumer desire for elegant wines that complement meals rather than powerful wines that trounce them. Key states and regions follow, organized from west to east, north to south.

UNITED STATES
WINE REGIONS

CANADA

Seattle
Puget Sound
Columbia
Walla Walla
Willamette Valley

Snake

San Francisco
North Coast AVA
Napa + Sonoma
Lodi
Central Coast AVA
Santa Cruz Mountains

Colorado

Denver

Rio Grande

Los Angeles

Austin

Mississippi

Missouri

Chicago

Washington DC

Nashville

Philadelphia

New York City
Finger Lakes
The Hamptons

Monticello AVA

Miami

MEXICO

ATLANTIC OCEAN

PACIFIC OCEAN

WASHINGTON STATE

Washington State is the second-highest producer in the United States (behind California). The state can be thought of in two distinct parts: west of the Cascade Range mountains (where it is generally cold, damp, and difficult for grape growing—accounting for 1 percent of the state's vines) and east of the Cascades, which is largely a desert. Nearly all of Washington's vineyards are east of the Cascades and bask in long, sunny days thanks to a rain shadow effect. Geologically, Washington has an exciting past, and even today it remains a volcanic hotbed. Soil throughout most of the state is a mix of loess and silty loam sands on a bedrock of basalt, with topsoils deposited by glacial melt known as the Missoula floods at the end of the last Ice Age. Yakima Valley, Red Mountain, and Walla Walla are the state's most important AVAs for wine, and red grapes that thrive in heat and sunshine get the lion's share of vineyard land. Merlot, Cabernet Sauvignon, and Syrah are the three dominant varieties.

Despite Washington's position as the second-largest producer in the United States, much of the state's production is controlled by big corporations. More than one-third of all vineyard land is owned and managed by Ste. Michelle Wine Estates, which was once a subsidiary of a tobacco company (Altria Group) before being sold to a private equity firm (TMI). At least 80 percent of grapes in the state are machine-harvested. Walla Walla, an AVA in the south, has emerged as the state's most quality-oriented AVA, with notable producers like Gramercy Cellars, Leonetti Cellar, and Andrew Will Winery. Its wines are usually characterized by dark, deeply jammy fruit bolstered by plenty of alcohol.

OREGON

Oregon's wine story begins as recently as the mid-1960s, when Diana and David Lett and Charles Coury traveled north from California to Oregon's Willamette Valley to plant Pinot Noir. Soon after, in 1987, Burgundy's esteemed Drouhin family purchased a winery in the Willamette Valley sub-AVA of Dundee Hills, legitimizing Oregon on a global scale, and the rest is history.

Oregon's best-known region is the Willamette Valley, located in the northern part of the state, bordering Washington. Temperatures are cool to moderate, and the Pacific Ocean is a major climatic influence. Cool breezes from the ocean ensure plenty of acidity, and Oregon's notorious rainy weather can lead to rot or dilute wines in difficult vintages. Soils vary but are generally a mix of sand, clay, and loam.

Pinot Noir is the state's most important grape and accounts for 70 percent of all vines planted. Oregon's expression has emerged as a lower-alcohol and earthier alternative to California Pinot Noir (known for bold fruit). Some of the most exciting growers in the United States are making wines here. Domaine Drouhin, Big

Table Farm, Bow & Arrow, Evening Land Vineyards, Fossil & Fawn, Belle Pente, and Walter Scott Wines have long been favorites, and Chosen Family Wines is a new favorite founded by NBA stars Channing Frye and Kevin Love. It is a beautiful example of wine doubling as a vehicle for connecting friends.

CALIFORNIA

California, responsible for more than 85 percent of wine production in the United States, has a vibrant viticultural history dating back to 1783, when Franciscan monks first produced wine from the Mission grape (aka País in Chile and Listán Prieto in the Canary Islands), first brought to Mexico by Spanish settlers who then traveled northward. California's wine scene enjoyed moderate success for almost two centuries, weathering phylloxera in the late nineteenth century, then Prohibition, and then World War II. It finally blossomed in the 1960s, thanks to a new generation of Napa Valley winemakers (Robert Mondavi, Joe Heitz, and Mike Grgich among them) mentored by a man named André Tchelistcheff, who had worked in France and brought technique to a nation that mostly preferred sweet, fortified bulk wine.

Today, California is a juggernaut. The state is divided into five major AVAs, or American Viticultural Areas—similar to France's AOPs, but limited to geographical boundaries, without any governance over grapes, farming practices, or production methods. These five major AVAs are North Coast, Central Coast, South Coast, San Francisco Bay, and the Sierra Foothills. Each of these five large AVAs include smaller, more specific AVAs within them as well.

The North Coast AVA, which includes Napa, Sonoma, Mendocino, and Marin counties, is California's fine wine epicenter. Cool air from the Pacific Ocean mitigates warmer temperatures inland, creating diurnal shifts between long, sunny days and cool nights. Terroir varies significantly within smaller AVAs (Napa

US LABELING REGULATIONS

The following list explains the percentage of grapes required for labeling by appellation.

· If labeled by country, state,* or county: 75 percent (*California and Oregon are exceptions to the rule: 100 percent of grapes have to come from these states if the state is listed on the label. Washington State is also an exception and requires 95 percent of fruit to come from within its borders.)

· If labeled by American Viticultural Area (AVA): 85 percent

· If labeled with a single vineyard: 95 percent

· If labeled by grape: the stated grape must make up at least 75 percent of the blend

Valley AVA, which is an AVA itself within the North Coast AVA, includes sixteen sub-AVAs alone, from Los Carneros in the south to Calistoga in the north). On the Cabernet Sauvignon front, producers like Diamond Creek Vineyards in the Diamond Mountain District AVA and Snowden Vineyards and Corison Winery in St. Helena have long been sommelier favorites on account of their dedication to thoughtful (generally organic) viticulture and lower-in-alcohol, terroir-driven wines. For a riper style, wineries like Harlan Estate, Screaming Eagle, and Colgin Cellars set the benchmark with Cabernet Sauvignons that established the blueprint for cult wineries with private mailing lists across the globe.

Napa is a region that becomes much more nuanced with a brief topographical understanding. Its terrain ranges from sea-level valley floors to mountain peaks. Calistoga, Rutherford, and Oakville are all in the valley, so vines benefit from warm days and silty valley-floor soil. Howell Mountain and Diamond Mountain are in the northern part of Napa, on opposite sides of the valley, where grapes get much riper because the slopes bask in warmer inland temperatures and produce ripe, juicy expressions of Cabernet. This is in stark contrast to the more southerly Mount Veeder, close to San Pablo Bay, which is regularly blasted with cool sea air and produces crisper wines with bright natural acidity.

West of Napa Valley lies Sonoma Valley AVA, responsible for 6 percent of California's wine production. Sonoma Valley's nineteen sub-AVAs include the Alexander Valley, Dry Creek Valley, Sonoma Coast, and the Russian River Valley. Pinot Noir and Chardonnay have become synonymous with this area, producing juicy, fruit-forward styles that are delicious. (Hirsch Vineyards makes benchmark examples.) Zinfandel thrives here, too.

Chardonnay is increasingly grown farther south in Santa Barbara County, which has a long history of viticulture (it was the site of one of California's original missions) but escaped modern attention until recently, when Richard Sanford and Michael Benedict rebelliously planted what is now the Sanford & Benedict Vineyard in 1971 and which formed the basis of the Sta. Rita Hills AVA, within Santa Ynez Valley AVA. Today, the Sanford & Benedict Vineyard is responsible for some of the most balanced Chardonnays and Pinot Noirs in the state. (They were told vines could never grow in the cold temperatures!)

Lodi AVA, east of Napa Valley, produces more than 20 percent of California's wine and is having a resurgence, thanks to its sandy soil and very old vines that are getting attention from talented young winemakers like Tegan Passalacqua of Sandlands Vineyards and Turley Vineyards and Morgan Twain-Peterson of Bedrock Wine Co., as well as from journalists like Randy Caparoso. Their intellect and enthusiasm have started a movement to protect these old and historic vines.

NEW YORK

The Finger Lakes AVA has emerged as New York's most prominent, and there are some wineries making very good wines. Ukrainian-born Dr. Konstantin Frank first planted *Vitis vinifera* in the Finger Lakes area in 1962, and the winery is still producing terrific wines today. Organic viticulture in New York is challenging and rare, but the movement is beginning. Hermann J. Wiemer Vineyard, founded in 1979, is a pioneer in the Finger Lakes region and has long made compelling wines by anyone's standards. The winery has farmed organically since 2003, eschewing herbicides and synthetic farming practices, and it is in the process of converting some vineyard plots to biodynamics for Demeter certification (the global standard for biodynamics). Beyond Finger Lakes, vines are also grown on Long Island, which has two AVAs of its own, for the North Fork and South Fork. Channing Daughters in the Hamptons makes a wonderful lineup of interesting wines and a terrific selection of rosés!

CHAPTER 8

WINE'S JOURNEY TO YOU (AND WHAT TO DO WHEN IT ARRIVES)

Congratulations! Now you know how soil and climate, geography and geology, winemaker decisions, historical influence, and the grapes themselves affect the wine in your glass. This chapter aims to arm you with information and tools to navigate retail shelves and wine lists, choose wines for your weeknight dinner or special-occasion dinner party, and learn how to serve and store wine like a sommelier (if you care to do that at all!). We begin with a little background context for distribution in the United States, for which we can thank Prohibition and clever, if self-interested, mobsters like Al Capone. While distribution certainly isn't the sexiest part of wine, it's incredibly influential as it determines your access to what's available in your neighborhood. The pages that follow will hopefully empower you to choose wines that you're going to like at a store or a restaurant, to know when it's better to order a cocktail instead, and to answer the nuts-and-bolts questions that my non-industry friends and family have been asking since I chose wine as a career decades ago. Cheers to you—thank you for making it this far!

BUYING WINE IN THE UNITED STATES

While in some respects buying wine anywhere in the world is the same (choose a bottle, purchase it from the restaurant or retailer, then do what you'd like with it because now it's yours), every country has its own nuances. In France, even at the best retailers (like Caves Augé or Caves Legrand in Paris), there's usually a hidden stash of sought-after wines in the back that you can access only when you converse with the buyer, who then assesses your worthiness. In the United States, things are slightly less nuanced, and power and access are wielded in large part by the distributor.

AMERICA'S THREE-TIER SYSTEM

The Twenty-first Amendment to the US Constitution, ratified in 1933, repealed the Eighteenth Amendment, ended Prohibition in the United States, and implemented a mandatory three-tier system for liquor sale and distribution, essentially baking distributors into the wine and spirits ecosystem. According to the website of NABCA (National Alcohol Beverage Control Association), America's three-tier system is defined as "an hourglass-like structure with producers at the top, funneling down to the wholesalers in the middle where tracking of tainted alcohol products and excise alcohol tax collection occurs, and then fanning back out to retailers who get the product to the consumer." These three tiers include the winery, the distributor, and the retailer or restaurant who then sells the wine to you. The three-tier system mandates that no one involved in US restaurants or retail shops can have any connection to wineries or distributors, in what are known as the tied-house laws.

Distribution, the middle tier, emerged from organized crime during Prohibition. Mobsters had a lot of money and therefore influence in the early twentieth century, which they used to ensure that the concept of wholesalers or distributors remained in place in America after Prohibition ended. If you're wondering why a European winery can't just sell you wines directly, America's three-tier system is the reason. This system has been causing consternation among wineries and wine lovers for almost a century. Until recently, individual states were allowed to ban direct shipping from out-of-state wineries (while still allowing direct shipping from those wineries in-state), which the Supreme Court ruled unconstitutional in 2005.

Today distributors, or wholesalers, come in all shapes and sizes. Some of the big ones are still involved with organized crime—and some of the smaller ones are run by former sommeliers who moved from cities like New York, Chicago, or San Francisco to other markets, determined to bring the wines they love to the places they now call home. There are also midsize distributors and everything in between.

DISTRIBUTION IN THE UNITED STATES AND HOW IT IMPACTS YOU

The distribution system in the United States impacts your life in a few ways. On one hand, it affects the price you pay to drink your favorite bottle of wine or liquor. There are some middle people in between the winery and the bottle on your table, and every time that bottle changes hands (or transits across the country), the price increases. The winery sells wine to the importer (who is also sometimes the distributor—otherwise it sells to the distributor), and the distributor sells that bottle to a retail shop, a hotel, or a restaurant, who then sells it to you. That is one of the reasons wines can be more expensive in the middle of the country than on the coasts—shipping charges increase the price of your bottle.

On the other—and more important—hand, the presence of distributors affects your access and choice, and therefore your actual world of wine. As large national distributors continue to consolidate portfolios, their number of selections grows unwieldy—often upward of twenty-five thousand different wineries, not to mention all of the different wines each of those wineries make. One friend of mine who helped launch a wine portfolio for a large famous winery through a large

national distributor mapped out *eleven* different people involved (and therefore eleven layers of the system) between the winery and the end consumer. That is a very long game of telephone, in which the consumer is meant to learn about wine from a retailer who sells it, who is getting information ten times removed from its original source.

In theory, a sales rep is supposed to know and care about every wine in their portfolio. But knowing even a fraction of that is unrealistic when portfolios are so big. What tends to happen instead with many large distributors is that their sales representatives sell whatever wine(s) are on their quota (wines they have to sell in order to make their bonus) that month. Their bonus is contingent upon them selling those things at the expense of everything else. Wineries get their wines on quota if they offer the distributor enough cash incentives, essentially ensuring their wines continue to get pumped into the supply chain and onto shelves.

Within large distributors, large wine and liquor groups are prioritized because they can allocate sales and marketing dollars to pay sales reps to recommend their selections. What those portfolios lack in inspiration, they make up for in volume, infrastructure, and marketing budget. Therefore, those are the brands that line the most shelves. Are they great? I don't drink them. Are they everywhere? Yep, they are everywhere.

ALLOCATIONS

Truly great wines sell themselves, so when large distributors are lucky enough to have them in their portfolios, they are allocated to customers based on the amount of crappier wine in their portfolio those customers are also willing to buy. Large distributors leverage their allocated wines in order to sell exponential amounts of wines no one really wants.

When I wrote my first wine list at a restaurant on New York's Upper East Side, I was told I needed to sell through fifty cases of Matua New Zealand Sauvignon Blanc if I wanted to qualify for four bottles of Vincent Dauvissat Chablis, one of the great wineries in the world that probably does not know Matua Sauvignon Blanc even exists. I got a taste of what it meant to be courted by large and powerful national distributors who offered me extreme discounts, special trips, and a lot of "spend" at the restaurant (promises to come and spend money at my establishment if I committed to pouring a particular wine by the glass). While some restaurants and hotels are able to prioritize wine and invest in a sommelier or wine professional committed to building a quality wine program, a lot of them—especially post-COVID—are not. That means the local liquor rep ends up writing the list (from their own portfolio) in a bid to keep costs "efficient" for the restaurant.

Please take a moment of gratitude for your favorite restaurants and retailers that offer great selections, because stocking great wines (and fending off reps who are trying to sell bad ones) is a tiresome thing to do!

By contrast, small, quality-minded wineries often seek out smaller, quality-minded distributors, who in turn sell to high-integrity retail shops and ingredient-focused restaurants. Smaller distributors tend to operate very differently, supporting wines and beverages they actually drink and whose makers have values and stories they believe in. The quantities are much smaller, so you need to know where to look.

Wine stores, like distributors, come in all shapes and sizes, and they all have different roles to play. A national discount retailer's job is to offer you things you've heard of for a lower price. There is a lot of value in that! The wine shop you keep hearing about from your friends who love wine has a different job—to broaden your horizons, thrill you with new selections, and make sure you are stocked with something great for your upcoming dinner party. Those wine shops offer value of a different kind.

Generally speaking, great wine lists exist in markets where there is a wine culture—enough collective interest to invest in learning about, purchasing, and drinking great wines. These don't have to be expensive wines! But it is helpful to have a culture that knows and appreciates what those wines, inexpensive or not, are and cares enough to inspire importers and distributors to carry them.

Many great bottles these days are available nationwide if you know where to look (and who imports or distributes them). Regardless of where you live, there's always the internet. My dad, who lives in Denver, has amassed a remarkable cellar by purchasing online from his favorite retailers, many of whom are out of state. You can sign up for mailing lists, peruse their websites—or, better yet, if you're in the area, walk through the stores and ask questions. When they're on duty, wine people love to talk about wine.

WINE LISTS ARE NOT CREATED EQUAL (OR, HOW TO KNOW WHEN IT'S TIME TO ORDER A COCKTAIL)

The question I am asked most often by friends who do not work in wine is, "How do I navigate a wine list without embarrassing myself?" Luckily, there are some quick ways to assess if a wine list is worth your time, and if a bottle from it is likely to be a pleasurable addition to your education.

SUSS IT OUT CONVERSATIONALLY: If you're in a talkative mood, ask the server if there's a sommelier (*suh-mal-YAY*) on hand. If there is, terrific! It means someone in ownership cares enough to invest in an actual person whose job is to manage the beverage program. If there isn't (more and more likely post-COVID), ask a few more questions to understand who is doing the work. If you're lucky, perhaps it's an ambitious and curious server who writes the wine list on their day off. If not, perhaps it's the local liquor distributor rep, pushing through uninspired wines from his company's quota list. There are plenty of programs that inspire me to order a cocktail instead.

WINE LIST PRICING

A great restaurant wine program does not use formulaic pricing when marking up bottles. Simply taking the cost of a bottle and multiplying it by 2.75 or pricing a glass of wine as the cost of the bottle is one way to meet your numbers, but great sommeliers are editors and curators, not robots, and their lists will reflect that.

Great wine lists will always have underpriced gems, and you can find them by chatting up the sommelier and getting to know what's off the list (but in the cellar), and what bottles they think are particularly special. How does that work? The role of by-the-glass wines in the most thoughtful wine programs is essentially to subsidize the rest of the list. Your Pinot Grigio, rosé, and Sauvignon Blanc by the glass? Those are the workhorses, probably marked up five or six times their cost, because people are going to buy them no matter what, and those people are going to be happy to pay that price.

When I worked as a sommelier and wine director at a restaurant on Manhattan's Upper East Side, many guests requested wines from Bordeaux. I added a few good selections from various villages and at various price points, then marked them up seven times their bottle cost. The wines didn't just sell—they flew out of the cellar! The wines were well made, they were stored and served properly, and they bankrolled the rest of the wine list. This allowed for fun rare-glass pours (we opened a different magnum of Champagne every night and sold glasses at cost) and other underpriced bottles to exist. It was a healthy wine ecosystem in which everyone benefited.

BUYING FROM A WINE STORE

If you're shopping at a wine retailer or grocery store and wondering how to navigate the aisles, you have some options. As with wine lists at restaurants, the most important thing is to go to a spot that truly, deeply, emphatically cares about wine. Certainly somewhere near you has an interesting collection, and one of my favorite pieces of advice—irrespective of wine—is "start where you are."

If you're a "live in the moment" kind of person, ask the person at the shop what they're excited about. Ideally, give them a budget range and a data point or two. For example, "I recently had a great bottle of Emmerich Knoll Riesling from the Wachau, in Austria, that I loved. What can you recommend that's similar?" Or if you'd like to keep it more open-ended, you could say you would appreciate a recommendation for something hand-harvested, organically farmed, and great with the roast sea bass with tomatoes and capers that you're making tonight.

If you're a "plan in advance" kind of person, print out a great wine list (I still love Eleven Madison Park's) and use a handy website like Wine-searcher.com to find out which retailers in your area have particular bottles. (I've found plenty of wine stores this way, tracking down bottles like a truffle hound.)

Assuming the wine list is one you want to tackle, approach it as you would anything—with curiosity and an open mind. A good restaurant or wine shop should make you feel comfortable, not intimidated. Remember: No one in the world, not even the most esteemed professional, knows everything about wine! The person who is best able to help you find a great bottle is the person who created the list in the first place. Befriending the sommelier is your passport to a vinous adventure.

A good place to start is with whatever is most important to you. Do you have a budget of fifty dollars? A good wine list will have options at any price point. If you recently drank a wine from a region you enjoyed and want to go farther in that direction, by all means continue. Giving someone your budget, preferences, and intended meal should allow them to come up with a few solid suggestions for you. Many wine professionals are excited by this challenge, as the possibilities can be endless. You may not always get the perfect match, but you will always learn something. Be open to the process! Alternatively, you may not be in the mood for an adventure. You may just want a great bottle of Savennières, and you can ask for that! You get to drink what you want. It's your dinner, after all.

THE SILENT WAY: If you don't feel like chatting, the fastest way to understand the ethos of a program is to look at the Champagne selection. The big-name Grandes Marques—like Moët & Chandon, Champagne Pommery, Veuve Clicquot, and Dom Pérignon—are currently distributed by large liquor distributors. One distributor alone accounts for 22 percent of the entire Champagne export market. If you peruse the list and there's nary a grower in sight, this means the buyer is probably just building a list out of wines it tacks onto its liquor orders. The person in charge here is not making inspired choices, and they're looking out for their best interests, not yours.

Scan the Champagne list for small, excellent, and widely distributed grower-producers like Bérêche et Fils, Chartogne-Taillet, Pierre Péters, Larmandier-Bernier, Pascal Agrapart, or Jacques Selosse. If you see these names, someone at that establishment has made some effort. I'd be happy to drink one of those bottles and give the list a chance.

THE RESTAURANT WAY: Another surefire way to find a great bottle is to select a restaurant known for its wine program. When at least one of the owners is a wine lover or professional, you can be certain any bottle on the list landed there with intention. Restaurants like Frasca Food and Wine or Tavernetta, both in Colorado, are benchmarks, along with A16 in San Francisco and A.O.C. in Los Angeles, as well as Le Bernardin, COTE, Eleven Madison Park, and The Four Horsemen in New York City. Any place in the Danny Meyer or Daniel Boulud restaurant groups also has a reliably great program; not surprisingly, the owners love wine! The landscape changes too frequently for me to make a reliable list, but just know that any restaurant with an owner who truly loves or appreciates wine is going to be happy to answer any questions and ensure you feel great about whatever wine you choose.

FOOD AND WINE, TOGETHER

There's a lot of flexibility when it comes to finding a great pairing of food and wine; it's a calculus that also involves a good deal of your opinion. Or sometimes no opinion at all—just a love of a good bottle of wine and a desire to drink it with a meal you also love. When I worked at Daniel, one of New York City's most famous French chefs used to come in for dinner and order oysters with a bottle of red First Growth Bordeaux. To this day, you could not pay me to drink those together, but that's what he liked, and no one is questioning his palate.

I have a complicated relationship with pairings, probably because I spent several years focusing on them at Eleven Madison Park and Momofuku Ko, and also because composed pairings on a restaurant menu don't match the way I actually like to drink most of the time. (Far too many restaurants use them as a way to move through cases of whatever the previous sommelier bought on closeout or a whim.) On the rare occasion that I am willing to let someone else design my journey, I'll go to a trusted restaurant with a passionate and purposeful person at the helm, like Aldo Sohm at Le Bernardin or Raymond Trinh at Sixty Three Clinton on Manhattan's Lower East Side. If I'm cooking, I'll drink a spritz during dinner prep and then have a glass of wine, maybe two, with dinner. That said, I'm still following a set of mental guidelines honed over the years. They're below, should you find them helpful.

WHAT GROWS TOGETHER, GOES TOGETHER

You've probably heard this kitschy saying before. But guess what—it's often true! Prior to the internet and airplanes, people lived—and therefore ate and drank—regionally. Some pairings unlikely to be improved upon include a terrific Chianti Classico with a bowl of pasta all'Amatriciana or a plate of prosciutto, or a glass of briny, orange-zesty Galician Albariño with cured anchovies and pan con tomate. This is true of dishes anywhere with a wine culture that evolved alongside its cuisine.

WINE OF THE TIMES

In Hermitage, in France's Northern Rhône, Jean-Louis Chave, winemaker at the domaine his family has run since 1481, speaks of the change in cuisine over the past few generations and, subsequently, the shift in wine style trends. Hermitage Blanc, which Thomas Jefferson famously called "the first wine in the world, without a single exception," has almost no acidity. In turn, it was an exceptional counterpart to meals made largely with butter, an ingredient considered the ultimate luxury and delicacy in the region at the time.

White wines from Burgundy, which have become brighter in recent years, used to optimize for ripeness and richness, with vignerons harvesting as late as possible and aging wines in new oak, with plenty of bâtonnage (lees stirring) for extended periods of time. The sixteenth through nineteenth centuries produced golden, buttery wines with lots of weight and texture, a perfect match for rich dishes like foie gras terrine, oeufs en meurette (poached eggs resting in a sauce of red wine thickened with butter), and the ripe, unctuous cheese Époisses (allegedly Napoleon's favorite). But as gastronomic preferences have skewed lighter in recent years—with vegetables being more desired than offal on most people's dinner programs, and olive oil often replacing butter as a preferred fat—the wines have followed suit, evolving into leaner versions of their past selves.

PAIRING GUIDELINES

PAIR FOR TEXTURE: Texture is as important as flavor when you're finding the perfect food and wine match.

Light dishes, like a fresh salad or a lean protein such as oysters or fluke crudo, call for something light-bodied—a dry or barely-off-dry Mosel Riesling, or something from a cool climate, perhaps Chenin Blanc, Sauvignon Blanc, or Melon B. You don't want to overwhelm the dish! Light-bodied also means low in alcohol. Drinking something from a cool climate will help ensure your pairing stays balanced and your wine choice doesn't overwhelm the dish.

Creamy, luxurious foods call for creamy, luxurious wines. A bowl of clam chowder or a bisque requires a wine that can handle that weight. An Oregon Pinot Gris from a warm vintage could be terrific here, as would a white Rioja made from the delightfully waxy Viura grape, which then ages in oak for years and acquires additional texture. In short, seek out something rich in weight and almost absent in acidity.

Red meat is a very good match for wines high in tannins. Rib eye is a classic pairing with Cabernet Sauvignon, Nebbiolo, and Sangiovese because both red meat and high-tannin wines are structurally powerful in their own right. All three grapes produce high-tannin, high-alcohol wines, and they also all produce wines with a healthy dose of acidity (useful for cleansing the palate after all of that fat!). Tender lamb with Syrah is a home run every time, thanks to the grape's gentle tannins and herbaceous overtones. Of course, mix and match as you'd like, but Syrah has much gentler (and far fewer) tannins than Cabernet Sauvignon, Nebbiolo, and Sangiovese, and lamb is a more delicate meat, with a gamey flavor profile.

OPPOSITES ATTRACT (ACID AND BUBBLES, WITH FAT): On the other hand, especially with fried things, opposites attract! Champagne with French fries (or fried chicken, or fried anything, really) is hard to beat. The perfect counterpoint to a cold, high-acid effervescent sip of Champagne is a warm, crispy, salty, fatty bite of food. Acidity helps cleanse your palate, the warm and cold temperatures are delicious complements to each other, and the bubbles add an extra dimension and lift. While Champagne is always a good idea, any delicious—and chilled—sparkling wine will work. German Sekt from the right producers (Van Volxem and Peter Lauer are at the top of my list) could be terrific stand-ins, as would a good Crémant.

MATCHING FLAVORS: While I tend to think about matching textures first, matching like flavors is a good strategy, too. Below are a few general flavor groups along with the ways I think about pairings when I'm in charge of selecting wine for a meal.

CITRUS: This one offers a lot of opportunity for you to be creative, depending on the type of citrus and the type of dish. Pretty much every white wine involves some expression of citrus, so it's up to you here to determine what wine is going to match with the dish at hand. Serving roast chicken with preserved lemons? A weighty Chardonnay or a Spanish Godello could be compelling. If the dish in question is lighter and brighter, like razor clams with fresh lime, maybe play up the lime and go with a Riesling. Spaghetti alle vongole with lemon and parsley? Consider opting for a wine from a coastal region, like Falanghina from Campania, or a Pigato from Liguria—like fresh lemons and salt spray, right by the sea. Any dish featuring grapefruit is a strong contender for pairing with Sauvignon Blanc. You get the idea.

BUTTER: In a course that puts butter on a pedestal, seek out wines voluptuous enough to match the butter and dynamic enough to play homage to other flavors in the dish. With sole à la meunière, butter and lemon are the stars, so richness and acidity are both important. A white wine aged in oak, probably one that also underwent a malolactic fermentation (it will have some hints of butter, thanks to that diacetyl), is where my mind goes first—but you may have other ideas, and you should test them out! My default would be a Premier Cru Chablis or a Mâcon, depending on whether lemon is the star player (Chablis) or brown butter (Meursault or Mâcon), but a good Chardonnay from Oregon or Santa Barbara would be stellar, too. Tired of Chardonnay? How about a Friulano, which has plenty of palate weight and savory herbaceous notes!

MEAT AND GAME: Meat offers the opportunity to open a bottle of red wine with lots of tannins. We touched on this earlier in regards to texture (see page 215), but finding the right bottle is about flavor, too. Sangiovese from Tuscany is terrific with steak (home of the bistecca alla fiorentina, after all), thanks to its grippy tannins—but flavor-wise, it's phenomenal on account of all those herbaceous and sanguine notes. Gamey meats and birds like lamb, duck, squab, or venison are classic pairings with Northern Rhône Syrah and delicious with Mourvèdre (the grape that reminds Daniel Boulud of "blood and guts in the sun"). And roast chicken with Pinot Noir is a no-fail, tried-and-true match made in heaven.

One pairing I will never forget occurred during a two-night collaboration with the famed chef Massimo Bottura of three-Michelin-starred restaurant Osteria Francescana (which had just snagged the top spot on the list of The World's 50 Best Restaurants in 2016). He arrived at Momofuku Ko during the Grand Gelinaz! Shuffle, an event in which great chefs from around the world trade places with other top chefs and cook their own menus in a host restaurant for a couple of nights. Diners purchase tickets to a participating restaurant but have no idea which chef will be cooking until they arrive that night.

I remember quite a lot about that evening, but nothing more clearly than one particular pairing that Massimo assured me did not need my participation. "With these fresh baby vegetables and perfect little greens," he told me, holding up a miniature radish root and fluffing a few pieces of young arugula, "we will pour only one very cold glass of still mineral water." We served still Italian bottled water in a Zalto glass, without ice. His vision was to celebrate the purity and perfection of fresh baby vegetables and create a pairing that enhanced the dish without overpowering it. It was the pairing of the night—and a reminder to me to think outside the box and forever keep an open mind.

SALT: Some wines, especially those from grapes that grow by the ocean, are perceptibly salty. If you're having oysters, a briny Melon B is a winner every time. A lot of Austrian Grüner Veltliners are salty, too. But don't limit yourself! Think of these wines as a complement to your fleur de sel and reach for them when you're eating something that might benefit from some extra seasoning. Vermentino, Falanghina, Pigato, Mataòssu—there are so many to choose from, go and explore!

LEAFY GREENS AND HERB-HEAVY DISHES: Aromatic, bright, vegetal wines like Grüner Veltliner and Sauvignon Blanc are good go-tos when herbs or vegetables are prominent. A plate of asparagus, a platter of artichokes, or the almost-too-herby salad I make when I'm in charge of dinner are some dishes that spring to mind when I think of wines featuring these herbaceous grapes. Grüner Veltliner is loaded with rotundone (see page 34), which gives the wine a white-peppery note and, as a bonus, is going to act a bit like seasoning with every sip. Vermentino, Vinho Verde, Grüner Veltliner, and Sauvignon Blanc are all good places to start!

PAIRING FOOD WITH A SWEET OR OFF-DRY WINE: The word *sweet* scares everyone away, even though a lot of times off-dry wines (which means "a little bit sweet") are a perfect complement to food. Think of off-dry wines as the vinous counterpart to a perfectly balanced lemonade. Often, off-dry wines are accompanied by searing acidity, something I love, and I reach for them regularly if I'm eating a spicy dish like pad Thai or anything with Calabrian or Szechuan pepper. They're also useful complements to dishes that have a sweeter element or two, like a salad with peaches or tangerines, or—in the most obvious case—paired with dessert!

STORING AND SERVING WINE

The unfortunate truth is that it's possible to mess up a great bottle of wine by disrespecting it, either by storing or serving it improperly. The good news is that neither storing wine nor serving it is difficult; all it requires is a little attention.

STORING WINE

The temperature at which you store your wine is important because wine, like any organic entity, can spoil. I've read that 98 percent of Americans consume a bottle within one day of purchasing it; if that's you, a dedicated wine fridge isn't necessary. But if you are someone who keeps special bottles on a shelf or in a drawer, there are better options. And you deserve them!

In a perfect world, your wines would lie horizontally, in a dark, humidity-controlled place no warmer than 65 degrees and no cooler than 40 degrees (Fahrenheit). If you're starting to collect a few bottles, there are wine fridge options on the market as low as two or three hundred dollars, and a brand-name two-hundred-bottle EuroCave is less than two grand. I know, that's expensive! But it offsets the cost of accidentally ruining a great bottle you've been saving.

If you're not ready to buy a fridge specifically for wine, a good makeshift option is storing bottles horizontally in your refrigerator's vegetable drawer. This ensures the cork stays in contact with the wine, keeping the cork moist (and therefore effective). Dried-out corks allow oxygen into the bottle, which ultimately spoils the wine. If you're short on space, there are mini-wine fridge options for less than $150—that investment is a relatively small price to pay compared with the value of your bottles! Some retail stores will also let you store wines when you purchase them, especially once you've developed a good relationship (although then you will really need to plan in advance before deciding to drink them!).

There's no reason to keep your wine stored as low as 40 degrees Fahrenheit unless you're looking to store those bottles a very long time. Thanks to the relationship between temperature and oxygen, a colder cellar slows aging.

SERVING WINE

When it comes to serving wine, different bottles require different considerations. Drinking a bottle of Albariño that you bought last weekend and popped in your fridge? Just open and pour.

For fancier bottles that are bound for more of a special occasion, more effort is required. No matter the moment, make sure you have clean wineglasses. Nothing ruins a bottle like drinking from a smelly glass.

TEMPERATURE

A good rule of thumb is to serve still wines of any color between 55 and 65 degrees and sparkling wines between 45 and 55 degrees Fahrenheit (7 to 13 degrees Celsius). The more you taste, the more opinionated you'll become about what the "right temperature" means for you. I tend to like to drink my Champagne as cold as the ice bucket. It jolts my palate into focus and I just think it's most delicious that way. Plus, I'm not a fast drinker, so I still get to enjoy the multitude of aromas as they open up in the glass. Other people like to drink wine closer to ambient temperature so the flavors can bloom, like cheese, and that's their (and your) prerogative.

TOOLS TO SERVE WINE LIKE A SOMMELIER

The following tools will come in handy for serving wine. Some of my favorites are below, and while some are fancy, others are not.

You will need a **WINE KEY**! I am partial to Laguiole wine keys (corkscrews) because they are beautiful, effective, and practical. They also have exceptional knives for cutting foil in a clean line around the rim of the bottle. A pet peeve of mine has always been crappy knives that gnarl the foil and leave your bottle looking like a rabid raccoon got ahold of it first. Some sommeliers also like Code38 wine keys, which are metal and a little bit clinical for me, but they're still very high-quality and effective. (I have one, in fact. I just like my Laguiole better.)

219

WINE'S JOURNEY TO YOU

An **AH-SO** (a wine cork extractor) is helpful if you have older bottles or plan to open someone else's. The fancy name-brand version is called a Durand. It's a nifty two-part invention that is a marked improvement on the original ah-so and that allows pretty much anyone to open an old bottle without breaking the cork.

GLASSWARE is quite a rabbit hole. Thankfully we seem to have graduated from the era of wineglass companies trying to convince us we needed a specific glass for every kind of grape. (Who has storage space for that?) My personal preference is to have three sizes of glasses—one for crisp white wines and Champagne, another for red wines that benefit from some oxygen exposure (like a Cabernet Sauvignon or Merlot)—and another, in between, for everything else. If you can't be bothered, just get the last one. Zalto makes a great one (the "Universal"); although it's a splurge, it makes an excellent gift for wine lovers in your life. And while Zalto is the gold standard with its paper-thin sides, it also spawned a whole new generation of companies, like Gabriel-Glas, that are making high-quality options that look very similar. Finally, Roberto Conterno's Sensory Glass is incredible for bringing out the aromatics in fine wines, as well as for the way it rests in your hand when you hold it. The stem is shorter than other glasses, but it's more stable and easier to swirl without having to concentrate on holding the glass. When investing in glasses, make sure you like how they look, how they feel in your hand, how the glass feels when you take a sip, and, of course, if the wine shows well in them.

You'll want a **DECANTER**. Any glassware company also makes decanters. Less is more: avoid the ones with necks resembling brontosauruses (impractical, expensive) and look for something made of glass that you find beautiful. Zalto makes a lovely one (Carafe No. 25), Lobmeyr makes stunningly beautiful options if you have a budget to splurge, and a glass water pitcher works fine, too, if that's what you have on hand.

POLISHING GLASSES LIKE A PRO

If you've just invested in great wineglasses and you're afraid to use them because you don't want them to break, one key polishing technique is to make sure you polish the bowl of the glass separately from the base! A surefire way to snap that fragile stem is to twist the bowl and the base in opposite directions. A clean, soft, lightweight cotton napkin or dish towel works better than any advertised polishing cloths in my experience.

WHEN TO DECANT

There are three main reasons to decant a wine:

1. **TO AERATE A WINE (ADD OXYGEN) BEFORE DRINKING IT.** This works for white wines, younger red wines, and even Champagne. In white wines, it can help a "closed" wine open up and can help aromas like gunflint or onion skin (signs of reduction—see Wine Flaws, page 36) blow off. In this case, you would empty the entire contents of the bottle into the decanter and you don't have to be too careful while doing it. The point is to add oxygen, so letting the wine slosh out of the bottle into the decanter ferociously is okay.

2. **TO SEPARATE WINE FROM THE SEDIMENT AT THE BOTTOM OF THE BOTTLE.** High-tannin wines will produce quite a lot of sediment over time, as tannins link together and polymerize before becoming too heavy and precipitating down to the bottom. If you're planning to drink an older bottle of Cabernet Sauvignon or Nebbiolo, decanting is a great idea (unless you want big chunks of sediment in your glass).

   To decant for sediment, you'll need a candle (a cellphone flashlight works, too), a decanter, and a very steady hand. Hold the bottle over the light as you pour the wine into the decanter, peering with focus through the illuminated glass to make sure none of the sediment goes into the decanter. A good place to focus is the bottle's shoulder, as sediment begins to collect there. Peer relentlessly into the neck of the bottle, and as soon as you see sediment venture into the neck (where the cork was), stop decanting immediately.

3. **BECAUSE DECANTING IS FUN, AND YOU FEEL LIKE DOING IT!** It adds a bit of dramatic flair, especially in restaurants.

   If you are wondering about gadgets like wine aerators, I think they are gimmicky and a waste of money. If you are hankering for a new wine accessory, spend it on glassware or that wine fridge.

Storage and Serving Q&A

Below, in Q&A format, are questions friends have asked me through the years. I'm including them here in hopes that they will be helpful.

Q: What is the ideal serving temp for red versus white versus sparkling?

A: This is very much up to you. I tend to like drinking crisp, light-bodied white wines and sparkling wines quite cold (45 to 50 degrees Fahrenheit), and I'm happy to have fuller-bodied whites and light-bodied reds a bit warmer (55 degrees Fahrenheit). Fuller-bodied red wines are in their groove around the 60-degree mark for me. If you're not sure where to begin, these ranges are a good starting place, and know that your opinions may evolve as time goes on.

Q: Are magnums obnoxious or useful?

A: Magnums are great! They are big and festive. They hold twice the amount of liquid as a bottle of wine, but the cork is the same size. This means that the ratio of wine to oxygen is double that of a 750-ml bottle (in other words, there's twice as much wine in a magnum as in a standard bottle, but the same amount of oxygen), which makes a magnum the ideal vessel for aging wine, as the aging, or development, is halved.

Q: What is the purpose of half bottles? Are they any good?

A: These are perfect for nights when you want to drink less wine. By the same logic as the magnum but in reverse, these bottles tend to age or develop more quickly, given that the cork size remains the same but the bottle contains half the liquid. But this can be a good thing! For example, a bottle of Barolo that you may want to age for a decade will be quite delightful after just a few years.

222

Q: Do you have to decant? What about just opening a bottle and letting it breathe?

A: Decanting is important for older red wines that have sediment. But if you're not drinking a bottle that fits that description, decanting is not essential! If I'm planning to drink a young bottle of red wine, I'll just pour it directly into the glass and let it open up there.

Q: What if you don't want to decant an old bottle?

A: If you don't want to decant a very old bottle, select it two to three days in advance and stand it upright in your fridge or cellar. This will allow the sediment to settle to the bottom. Then, when you serve, pour the wine carefully, ideally never letting the bottle tilt farther than 45 degrees (to be as gentle as possible and not mess with the sediment!). A bottle is not a snow globe.

Q: What is double decanting and when is the right time to do it?

A: This refers to decanting wine off of the sediment and into a decanter, tossing the sediment down the drain, rinsing out the bottle with water a couple of times (until the water runs clear), then pouring the decanted wine back into the original bottle. You might do this if you are serving wine at dinner, and you'd like to be able to show off the label during service. You might do this if you are going over to someone's house and want to bring a bottle of wine to share, and to make sure it is sound—and decanted—in advance. Thoughtfully, this ensures you can taste and prepare the wine beforehand, leaving no complicated work for the host. Plus, there's often no better vessel in which to serve the wine than the original bottle.

CHAPTER 9

LAST SIP

Our thoughts, feelings, processes, and unconscious beliefs
have an energy that is hidden in the work. This unseen, unmea-
surable force gives each piece its magnetism. A completed project
is only made up of our intention and our experiments around it.
Remove intention and all that's left is the ornamental shell.
—Rick Rubin, *The Creative Act: A Way of Being*

If we think of wine as art, the intention—of everyone involved along the way—is an essential element in how that wine shows up in our glass. As art, its beauty and resonance are contingent upon everyone who played a role in crafting it: from the workers who tended grapes, harvested them, crushed them, and cleaned the tanks, to the winemaker whose vision helped transform the grapes into the finished wine.

The fun part, for you, gets to be determining whose philosophy and intention match your interest and curiosity as you go forth. You may be inspired to dig in and learn more. Or you may not! If your curiosity is piqued, below are a few ideas for how to follow your interest and even add a bit of structure to your approach, if that's your thing.

WHERE TO GO FROM HERE?

Okay, you love wine and want to learn more. What's next? Well, it depends what you're looking for. While taking a course can be a helpful credential if you're look-ing for a job in wine, you can learn at least as much on your own, if you're armed with a little self-discipline.

Start by following your curiosity. Learning about wine should feel like a joy, not a chore. If you drank a wine from the Canary Islands with a friend and loved it, take a photo of the label and google the producer and region. Try other wines from the region and the producer when you have the opportunity. If you are a person who likes to experiment, and you want to be a student of wine, take notes in a journal, because your brain processes and locks in information differently when you physically write it out. Snap photos of bottles you enjoy, and create a folder on your phone for wines you liked (and, for that matter, wines that you didn't). If you are into podcasts, Levi Dalton's *I'll Drink to That! Wine Talk* is great, featur-ing thoughtful interviews with winemakers, sommeliers, and importers alike.

If you prefer a more structured approach, the UK-based Wine & Spirit Education Trust is an internationally recognized organization with a thorough curriculum. The Court of Master Sommeliers is another option that will lead to a terrific community of wine lovers on a similar journey.

One of the most effective ways you can learn more about wine is to download a world-renowned wine list from the internet, print it out, and research a few wines each day. As I write this, the restaurant Eleven Madison Park is a great place to start, and it publishes its wine list online. Le Bernardin, Aldo Sohm Wine Bar, and Chambers, also in New York, and Frasca Food and Wine, in Boulder, Colorado, are other restaurants with renowned wine programs. Depending how busy you are and how quickly you're looking to learn, set aside a bit of time and treat this like meditating, or exercise, or even brushing your teeth. Be consistent! Pick a wine a day and commit to ten to fifteen minutes of focused attention. Most wineries today have informative websites, as do most distributors that import and distribute those wines. Or if you're not sure where to start, go into a restaurant or retailer that's known for its wine program (every town has one), and ask someone who seems to have a passion for what they're excited about on the list and why. If you can spend ten minutes a day for one month focused on wine that interests you, you'll be shocked at how the world of wine starts to make more sense with each passing day.

If you're feeling inspired after reading these pages, you can create a "wine one-sheet" for yourself. I used to make one of these for every wine we poured by the glass for our team at every program I oversaw, so everyone had a good handle on the different offerings (thanks for the original framework, John Ragan!). On a sheet of paper, write the name of the wine at the top, with the vital stats—like grape, vintage, region, country—right after that. Then jot down a bit (a few sentences to a few paragraphs depending on your mood, available time, and level of ambition) about the wine's history, soil, climate, producer—whatever stands out to you. Then, while sipping on a glass, pick five words or phrases and write them at the bottom of the page to jog your memory when you're reviewing your notes, or just to help your newfound knowledge sink in. (For example, I just dug up an old Momofuku wine one-sheet for a 2010 Domaine Albert Boxler Pinot Gris from Alsace. My five boldfaced phrases were *Iconic estate; 300-year-old family winery; organic; ripe stone fruit; clementines!* See—now I'm thirsty.)

By all means, taste what you're studying! Sip on a glass of what you're researching while you're researching it. When you do this, you are engaging all parts of your brain, creating a multisensory experience that will stay ingrained in you.

227

IN VINO VERITAS

All I know is that I know nothing.
—Socrates

Several years ago, I made a spontaneous visit to Montalcino, Italy, to taste with the winemaker Stella di Campalto. Stella makes wines I have always loved and found somehow more elegant, more Burgundian, than other wines made from Sangiovese Grosso, the Brunello di Montalcino clone. (No offense, Sangiovese Grosso. Hers are just extra special.) It was autumn and the harvest had just finished—grape leaves reddened and yellowed on vines. Wines burbled, fermenting in their wooden vats. And I heard monks chanting deep in Stella's cellar.

At first, I thought I was imagining things, but then I asked her about these voices. Stella shared that, during fermentation, she plays a CD of local Gregorian monks from the nearby Abbey of Sant'Antimo, chanting on repeat. One year, she didn't play the chants, and her fermentations got stuck. Now she does it annually. Whether it's coincidental or there's something bigger at play, her fermentations go smoothly when she plays the chanting monks. It has become a permanent part of élevage.

A few hours after that conversation, I laced up my running shoes and ran down to the abbey, which I'd long heard of and read about but had never thought to visit. Those chants were alluring, like Stella's wines, and I couldn't stop thinking about them. When I arrived at the church, a sign caught my eye. It explained the abbey's history, including its founding centuries ago by a group of Burgundian monks who fled France and ended up here in a part of Tuscany, near Montalcino.

I could not help but draw a conclusion that these ancient chants from Gregorian monks somehow link Stella di Campalto's wines to Burgundy. It was palpable and clear—and unbelievable at the same time. It was exactly the sort of mystery that makes great wine so special. No other wine from Montalcino behaves as elegantly as hers do. There's just something Burgundian in her wines.

I found this particular revelation so full of mystery and rife with questions to unpack. And I became very curious about music and winemaking. After a bit of research, I discovered that other great wineries use music in different ways, pointing toward the notion that there is something bigger than wine at play, bigger than all of us, that we can't quite see. Domaine Huet in Vouvray (in France's

Loire Valley) has found a way to stop the spread of a vine disease called esca (a fungus that infects the inside of grapevines) by recording the frequency at which esca grows, then reversing that sound and blasting it through loudspeakers to the vines at dawn and dusk, the times of day when esca is typically most active. They've all but stopped esca in that vineyard and are planning to roll this out to other vineyards. Domaine Comte Abbatucci, in southern Corsica, plays traditional Corsican polyphonic music to the vines over loudspeakers installed in the vineyards—an act of love and tradition meant to soothe the vines.

The phrase *In vino veritas* ("In wine, there is truth") still holds power. Wine can help uncover truths that we can't see with our eyes or even explain with our words, but we can feel them, intuitively, like the wind. In our quest for wines that excite and inspire us and bring us joy, we learn to regain our curiosity and our childlike senses of wonder, discovery, and the ability to live in the present. I'd argue there is no greater truth than that! I hope this book brings you closer to wines you love and people who inspire you, and that you feel smarter than you did when you picked it up. Cheers to you and your journey, wherever it may lead.

APPENDIX

FAVORITE PRODUCERS, BY REGION

The list below is a selection of wines and winemakers that make bottles I love, seek out, and purchase. It's not exhaustive (apologies in advance to any and everyone who should be on this list but isn't). It's also not a list that pays homage equally to the wines of every country, because that is not the way I drink wine. I hope that it's a helpful starting point for you if you find your palate resonates with mine.

ARGENTINA

Bodega Catena Zapata

Bodega Chacra

Bodega Noemía

Susana Balbo

AUSTRALIA

By Farr

Cullen

Leeuwin Estate

Moss Wood

Ochota Barrels

Vasse Felix

AUSTRIA

Bründlmayer

F.X. Pichler

Hirsch Vineyards

Moric

Prager

Rudi Pichler

Schloss Gobelsburg

Sohm & Kracher

Veyder-Malberg

Weingut Knoll

CHILE

Casa Marin

Jarilla Wines

Louis-Antoine Lyut

Pedro Parra y Familia

FRANCE: ALSACE

Domaine Albert Boxler

Domaine Marcel Deiss

Trimbach

FRANCE: BEAUJOLAIS

Domaine Chapel

Domaine Lapierre

Domaine Mee Godard

Jean Foillard

Julien Sunier

Yann Bertrand

Yvon Métras

FRANCE: BORDEAUX

Château Bourgneuf

Château Cheval Blanc

Château Cos d'Estournel

Château Ducru-Beaucaillou

Château d'Yquem

Château Lafleur

Château Léoville Las Cases

Château Lynch-Bages

Château Pontet-Canet

Pétrus

FRANCE: BURGUNDY

Alice et Olivier De Moor

Benjamin Leroux

Domaine Anne et Pierre Boisson

Domaine Armand Rousseau

Domaine Bachelet

Domaine Chanterêves

Domaine Coche-Dury

Domaine de Cassiopée

Domaine de la Romanée-Conti

Domaine de Montille

Domaine de Villaine

Domaine des Comtes
Lafon

Domaine des Croix

Domaine Dominique
Lafon

Domaine du Comte
Liger-Belair

Domaine Dujac

Domaine Eleni &
Edouard Vocoret

Domaine Fornerol

Domaine
Génot-Boulanger

Domaine Georges
Mugneret-Gibourg

Domaine Georges
Roumier

Domaine Jacques-
Frédéric Mugnier

Domaine Leflaive

Domaine Merlin

Domaine Pierre-Yves
Colin-Morey

Domaine Raveneau

Domaine Roulot

Domaine Simon Bize &
Fils

Domaine Sylvain
Cathiard & Fils

Les Horées

Moreau-Nadet

René et Vincent
Dauvissat

FRANCE: CHAMPAGNE

Bérêche et Fils

Champagne Cédric
Bouchard/Roses de
Jeanne

Champagne
Chartogne-Taillet

Champagne
Dhondt-Grellet

Champagne Jacques
Selosse

Champagne JM Sélèque

Champagne Krug

Champagne La Closerie

Champagne Pascal
Agrapart

Champagne Pierre Péters

Champagne Savart

Champagne Suenen

Champagne Ulysse
Collin

FRANCE: CORSICA

Clos Venturi

Domaine Comte
Abbatucci

Yves Leccia

FRANCE: JURA

Bénédicte et Stéphane
Tissot

Domaine du Pélican

Jean-François Ganevat

Maison Pierre Overnoy

FRANCE: LOIRE VALLEY

Catherine & Pierre
Breton

Clos Rougeard

Domaine de Bellivière

Domaine du Collier

Domaine Huet

Domaine Vacheron

Edmond Vatan

Jean Baptiste Hardy

Pascal Cotat

Stéphane Bernaudeau

Thibault Boudignon

FRANCE: PROVENCE & SOUTHERN FRANCE

Château de Pibarnon

Domaine de Trévallon

Domaine Tempier

L'Anglore

Le Roc des Anges

FRANCE: RHÔNE VALLEY (NORTHERN)

Domaine Alain Graillot

Domaine Clape

Domaine Jean-Louis
Chave

Domaine Romaneaux-Destezet (Hervé Souhaut)

Domaine Vincent Paris

Franck Balthazar

J.L. Chave Sélection

Monier Perréol

Pierre Gonon

Thierry Allemand

FRANCE: RHÔNE VALLEY (SOUTHERN)

Château de Fonsalette

Château Rayas

GERMANY: BADEN & AHR

Enderle & Moll

Meyer-Näkel

GERMANY: MOSEL

Egon Müller

Julian Haart

Peter Lauer

Selbach Oster

Stein

Van Volxem

Weiser-Künstler

GERMANY: NAHE

Dönnhoff

Emrich-Schönleber

GERMANY: RHEINGAU

Eva Fricke

J. B. Becker

Weingut Leitz

GERMANY: RHEINHESSEN

Keller

Weingut Wittmann

HUNGARY

Royal Tokaji

ITALY: ABRUZZO & LE MARCHE

La Distesa

Valentini

ITALY: CAMPANIA

Marisa Cuomo

ITALY: FRIULI-VENEZIA GIULIA

Borgo del Tiglio

Miani

Radikon

Venica & Venica

Vignai da Duline

ITALY: LIGURIA

Bruna

Ottaviano Lambruschi

Punta Crena

ITALY: PIEDMONT

Bruno Giacosa

Cantina Giacomo Conterno

Cascina Penna-Currado

G.D. Vajra

Giuseppe Rinaldi

Produttori del Barbaresco

Vietti

ITALY: SICILY

Benanti

COS

Eolia

Girolama Russo

Marco de Bartoli

Occhipinti

ITALY: TRENTINO-ALTO ADIGE

Elisabetta Foradori

ITALY: TUSCANY

Isole e Olena

Monteraponi

Poggio di Sotto

Soldera

Stella di Campalto

NEW ZEALAND

Felton Road

Rippon Vineyard

SOUTH AFRICA

A.A. Badenhorst

Craven Wines

Hamilton Russell Vineyards

Klein Constantia

Rebel Rebel Wines

The Sadie Family Wines

SPAIN

Bodegas Hidalgo La Gitana

Clos Erasmus

Clos Mogador

Comando G

CVNE

Envínate

Equipo Navazos

D'Oliveira

Finca Dofí

Guímaro

La Rioja Alta

Nanclares y Prieto Viticultores

Pingus

R. López de Heredia

USA: CALIFORNIA

A Tribute to Grace

Arnot-Roberts

Ashes & Diamonds

Bedrock Wine Co.

Calera

Clos Saron

Collate

Cruse Wine Co.

Dalla Valle

Diamond Creek Vineyards

Domaine Eden

Emme Wines

Hirsch Vineyards

Jolie-Laide

Littorai

Martha Stoumen

Massican

Matthiasson

Pax Wines

Phelan Farm

Renaissance

Ridge Vineyards

Sandhi

Sandlands

Snowden Cousins

Snowden Vineyards

Stolpman Vineyards

Turley Wine Cellars

USA: NEW YORK

Boundary Breaks

Channing Daughters

Dr. Konstantin Frank

Hermann J. Wiemer Vineyard

Terrassen

USA: OREGON

Belle Pente

Big Table Farm

Bow & Arrow

Chosen Family

Domaine Drouhin

Evening Land Vineyards

Fossil & Fawn

Lingua Franca

Walter Scott

USA: WASHINGTON

Andrew Will

Gramercy Cellars

Leonetti Cellar

Pollard Vineyard

SOME FAVORITE IMPORTERS AND DISTRIBUTORS

BECKY WASSERMAN & CO.: Founded by the late, legendary wine expert Becky Wasserman, this portfolio includes superstars at all price points making high-quality, soulful, terroir-driven wines. www.beckywasserman.com

BOWLER: David Bowler founded the company two decades ago, and it's grown to one of the best portfolios in the country. I've worked with them since writing my first mini–wine list for the terrace at the New York Plaza Hotel. In New York, Bowler includes the Louis/Dressner portfolio and so many others. I'm proud to say they now also distribute RAMONA in New York, New Jersey, and Pennsylvania. www.bowlerwine.com

GRAND CRU SELECTIONS: Known for exceptional-quality wines, notably from France, Italy, and the United States. Grand Cru's Burgundy and Champagne portfolios are particularly noteworthy. Note: Grand Cru is owned by Robert, my husband! While many of my favorite wines and wineries are now part of his portfolio, that has happened over the past few years (and in many cases, during my writing of this book) and is a testament to his commitment to quality, integrity, and incredible relationships earned through trust over the course of decades. www.grandcruselections.com

JOSÉ PASTOR SELECTIONS: A selection of exceptional, mostly Spanish wines that helped introduce America to Spain's new generation. These are some of my favorites in the world. www.josepastorselections.com

KERMIT LYNCH WINE MERCHANT: One of the original great portfolios for wines throughout Italy and France, with broad distribution. www.kermitlynch.com

LOUIS/DRESSNER SELECTIONS: The original "natural wine" portfolio, and still one of the best. Expect organic farming and bright, delicious, minimal-intervention wines from, primarily, Italy and France. www.louisdressner.com

VOM BODEN: Stephen Bitterolf's portfolio of extraordinary German wines is always a win. Stephen is a brilliant student of history with an insatiable curiosity and passion for Riesling. He has earned the respect and trust of all of Germany's great producers and is responsible for importing my favorite German wines. Riesling is the portfolio's strongest suit, but it also includes lesser-known white grapes as well as some spectacular Pinot Noirs from the Baden producer Enderle & Moll. www.vomboden.com

BOOKS TO CONSIDER

American Vintage: The Rise of American Wine by Paul Lukacs. One of my favorite books, period. I even chose it for my book club, as it's nonfiction that reads like a novel. It's very well researched and packed with terrific context about America's wine history from the beginning (Jefferson's failed attempts at Monticello) up to the present day.

Champagne: The Essential Guide to the Wines, Producers, and Terroirs of the Iconic Region by Peter Liem.

The Dirty Guide to Wine: Following Flavors from Ground to Glass by Alice Feiring with Pascaline Lepeltier. A great deep dive into the way soil impacts the taste (or "thumbprint") of a wine.

Flawless: Understanding Faults in Wine by Jamie Goode. A terrific guide for anyone curious about faults and how they manifest in your glass.

Inside Burgundy by Jasper Morris. A thorough and essential tome if you are falling in love with Burgundy.

Italian Wine: The History, Regions, and Grapes of an Iconic Wine by Shelley Lindgren and Kate Leahy.

Napa Valley Then & Now by Kelli A. White.

The New France: A Complete Guide to Contemporary French Wine by Andrew Jefford. While published in 2006, nearly two decades ago, Jefford's observations and insights are relevant and detailed, and he profiles producers who set the benchmarks for great wines today.

The New French Wine: Redefining the World's Greatest Wine Culture by Jon Bonné.

One Thousand Vines: A New Way to Understand Wine by Pascaline Lepeltier. A monumental book that blends Pascaline's probing mind and poetic prose with her philosophy background and covers details large and small with rigor and humility.

The Oxford Companion to Wine, 5th ed., edited by Julia Harding MW, Jancis Robinson OBE MW, and Tara Q. Thomas.

Secrets of the Sommeliers: How to Think and Drink Like the World's Top Wine Professionals by Rajat Parr and Jordan Mackay. This book is a terrific crash course on learning to think like a sommelier, written by the best of the best.

Vignette: Stories of Life & Wine in 100 Bottles by Jane Lopes. An incredibly thorough overview of the wine world through Jane's narrative. A love story of her own life, and an invaluable resource for anyone studying for an exam.

Vino: The Essential Guide to Real Italian Wine by Joe Campanale with Joshua David Stein.

Wine Simple: A Totally Approachable Guide by a World-Class Sommelier by Aldo Sohm and Christine Muhlke. This is a terrific resource that breaks complicated themes down into really smart, bite-size pieces of information.

The World Atlas of Wine by Hugh Johnson and Jancis Robinson. An essential overview that maps out the world of wine. If I were told I had to pick one desert-island wine book for the rest of my life, this would be it.

SOURCES

Alexander, Dean. "Understanding the Terroir of Burgundy, Part 1.1, Limestone Formation." *Diary of a Winebuyer* (blog), December 20, 2014. https://diaryofa winebuyer.wordpress.com/2014/12/20/understanding-the-terroir-of-burgundy/.

Asimov, Eric. "It's Time to Put the Noble Grapes in Their Place." *New York Times,* November 19, 2020.

The Australian Wine Research Institute. *Winemaking Fact Sheet: Using Malolactic Fermentation to Modulate Wine Style.* Urrbrae, SA: AWRI, updated August 2020. https://www.awri.com.au/wp-content/uploads/mlf_modulation _AWRI_fact_sheet.pdf.

B., Ilenia. "The Growing Popularity of Moscato d'Asti in the United States." *xtraWine Blog,* February 23, 2017. https://blog.xtrawine.com/en/the-growing -popularity-of-moscato-dasti-in-the-united-states/.

Biodynamic Demeter Alliance. "Biodynamic Principles and Practices." Accessed February 29, 2024. https://www.biodynamics.com/biodynamic -principles-and-practices.

Breitmaier, Eberhard. *Terpenes: Flavors, Fragrances, Pharmaca, Pheromones.* Weinheim, Germany: Wiley-VCH, 2006.

Brochart, Nans, and Brice Amato. *Rosé Wines World Tracking Report Summary: Key Figures 2021.* Provence Wine Council (CIVP) and FranceAgriMer, June 2023.

Bush, Zach. "Glyphosate + Toxins." *ZBMD Blog.* Accessed October 15, 2023. https://zachbushmd.com/gmo/glyphosate-toxins/.

Caparoso, Randy. "The Use of Varietal as a Term, Its History and Passing Practicality." *Coffee Shop Blog,* Lodi Wine Growers, March 8, 2021. https://lodigrowers.com/the-use-of-varietal-as-a-term-its-history-and -passing-practicality/.

Capstone California. "Wine Law & Official Classifications." California Wine Institute. Accessed October 15, 2023. https://capstonecalifornia.com/study -guides/regions/california/california/wine_law.

Clarity Wine Shop. "(ga)May 2023: An Examination of France's Bastard Grape." Accessed February 29, 2024. https://claritywineshop.com/blogs/wine-club /gamay-2023.

The Court of Master Sommeliers Americas. *Court of Master Sommeliers: Deductive Tasting Grid.* March 2022. https://cdn.courtofmastersommeliers.org /uploads/2022/04/Deductive-Tasting-Grid-March-2022.pdf.

Cristaldi, Jonathan. "10 Types of Wine Barrels Winemakers Love." *Food & Wine*, May 3, 2018. https://www.foodandwine.com/wine/wine-barrels-oak.

Editors of *Wine Enthusiast*. "The Beautiful Bounty of Botrytized Wines." *Wine Enthusiast*, updated May 4, 2023. https://www.wineenthusiast.com/culture/wine/the-beautiful-bounty-of-botrytized-wines/.

Eligon, John. "Exploring South Africa's Black Wine Scene." *New York Times*, February 15, 2023.

EQUALITAS. "Sustainability 'Made in Italy' for the World of Wine." Accessed October 18, 2023. https://www.equalitas.it/en/mission/.

Fisher, Rick. "The Wine Quality System of Spain." Wine Scholar Guild, November 18, 2019. https://www.winescholarguild.com/blog/the-quality-wine-system-of-spain.

Goode, Jamie. "The Science of Flor: What's That Growing on My Wine?" *Wine Anorak*, June 5, 2020. https://wineanorak.com/2020/06/05/the-science-of-flor-whats-that-growing-on-my-wine/#main.

International Organisation of Vine and Wine. *State of the World Vine and Wine Sector 2021*, April 2022. https://www.oiv.int/sites/default/files/documents/eng-state-of-the-world-vine-and-wine-sector-april-2022-v6_0.pdf.

Kane, Joe. "Understanding the Modern German Wine Landscape." GuildSomm, January 20, 2023. https://www.guildsomm.com/public_content/features/articles/b/joe-kane/posts/modern-german-wine-landscape.

Karlsson, Per and Britt. "Cabernet Sauvignon: The World's Most Planted Grape Variety." *Forbes*, June 22, 2020.

Khazan, Olga. "How France Became So Good at Wine." *Atlantic*, June 3, 2013.

Kornei, Katherine. "600 Years of Grape Harvests Document 20th Century Climate Change." *Eos* (September 27, 2019). https://doi.org/10.1029/2019EO134355.

Langley, Liz. "Do Animals Get Drunk?" *National Geographic*, November 21, 2015.

Lemelson-MIT Program. "Louis Pasteur: Pasteurization." Massachusetts Institute of Technology. Accessed October 18, 2023. https://lemelson.mit.edu/resources/louis-pasteur.

Luyten, Ruben. "Sherry Types." *SherryNotes*. Accessed October 18, 2023. https://www.sherrynotes.com/sherry-wine-types/.

Made Up in Britain. "Sparkling Wine." Updated August 8, 2021. https://madeupinbritain.uk/Sparkling_Wine#hn_Christopher_Merret_1662.

Mason, Olivia. "Shiraz: The Story of an Australian Legend." *Decanter,* December 2, 2019. https://www.decanter.com/sponsored/shiraz-story-australian -legend-427784/.

MasterClass. "The Complete Guide to Cava, Spain's Sparkling Wine." Updated September 28, 2021. https://www.masterclass.com/articles/the-complete-guide -to-cava-spains-sparkling-wine.

McKirdy, Tim. "All the Ways Wood Affects Your Wine, Explained." *VinePair,* September 5, 2018. https://vinepair.com/articles/wood-wine-aging-oak/.

McNicoll, Arion. "Commandaria: The Oldest Wine in the World?" CNN, updated December 13, 2013.

Nigro, Dana. "U.S. Supreme Court Overturns Wine-Shipping Bans." *Wine Spectator,* May 16, 2005. https://www.winespectator.com/articles/us-supreme -court-overturns-wine-shipping-bans-2543.

Nye, Bill. *Bill Nye the Science Guy.* Season 3, episode 4, "Rocks and Soil." Aired February 3, 1995, on Disney Channel. https://www.youtube.com/watch?v=N-K _CHgcazI.

Pearce, Kim and Michelle Gadd. "Sugar & Wine." *Organic Wine*, October 19, 2022. https://www.organicwine.com.au/sugar-and-wine.

Peck, Harry Thurston, ed. *Harper's Dictionary of Classical Literature and Antiquities.* Vol 3. New York: Harper and Brothers, 189.

Poniewozik, James. "The Best and Worst Super Bowl Commercials of 2010." *TIME*, February 7, 2010.

Poore, Benjamin Perley. *Perley's Reminiscences of Sixty Years in the National Metropolis*. Vol. 1. Philadelphia: Hubbard Brothers, 1886.

Robinson, Jancis. "Autolysis." Accessed December 5, 2024. https://www .jancisrobinson.com/ocw/detail/autolysis.

Robinson, Jancis. "Botrytis." Accessed December 5, 2024. https://www .jancisrobinson.com/ocw/detail/botrytis.

Robinson, Jancis. "Castilla y León." Accessed December 5, 2024. https://www .jancisrobinson.com/ocw/detail/castilla-y-leon.

Robinson, Jancis. "Eiswein." Accessed December 5, 2024. https://www .jancisrobinson.com/ocw/detail/eiswein.

Robinson, Jancis. "Merlot." Accessed December 5, 2024. https://www .jancisrobinson.com/ocw/detail/merlot.

Robinson, Jancis. "New World." Accessed December 5, 2024. https://www
.jancisrobinson.com/ocw/detail/new-world.

Robinson, Jancis. "Pinot Noir." Accessed October 15, 2023. https://www
.jancisrobinson.com/learn/grape-varieties/red/pinot-noir.

Robinson, Jancis. "Port." Accessed December 5, 2024. https://www
.jancisrobinson.com/ocw/detail/port.

Robinson, Jancis. "Prädikat." Accessed December 5, 2024. https://www
.jancisrobinson.com/ocw/detail/pradikat.

Robinson, Jancis. "Sauternes." Accessed December 5, 2024. https://www
.jancisrobinson.com/ocw/detail/sauternes.

Robinson, Jancis. "South Africa." Accessed December 5, 2024. https://www
.jancisrobinson.com/ocw/detail/south-africa.

Robinson, Jancis, Julia Harding, and Tara Q. Thomas, eds. *The Oxford
Companion to Wine*, 5th ed. Oxford: Oxford University Press, 2023.

Russan, Alex. "The Science of Terpenes and Isoprenoids." *SevenFifty Daily*,
May 25, 2020. https://daily.sevenfifty.com/the-science-of-terpenes-and
-isoprenoids/.

Sarl Fine Wine Experts. "Burgundy: A Brief History." Accessed October 15,
2023. https://www.finewineexperts.fr/wine-data/burgundy-region-2
/burgundy-a-brief-history/.

Schmidt, Michael. "The End of the Grosslage." Jancis Robinson, May 19, 2021.
https://www.jancisrobinson.com/articles/end-grosslage.

Schoonmaker, Frank. *Frank Schoonmaker's Encyclopedia of Wine.* New York:
Hastings House, 1978.

Sereni, Anthony, Quynh Phan, James Osborne, and Elizabeth Tomasino.
"Impact of the Timing and Temperature of Malolactic Fermentation on the
Aroma Composition and Mouthfeel Properties of Chardonnay Wine." *Foods*
9, no. 6 (June 18, 2020): 802. https://doi.org/10.3390/foods9060802.

Temple, Joseph. "8 Footnotes on Thomas Jefferson's Passion for Wine." The
International Wine & Food Society, December 5, 2014. https://blog.iwfs
.org/2014/12/8-footnotes-on-thomas-jeffersons-passion-for-wine/.

Thach, Liz. "Climate Change Propels France to #1 Largest Global Wine
Producer in 2023." *Forbes*, November 11, 2023.

Thach, Liz, and Angelo Camillo. "A Snapshot of the American Wine Consumer in 2018." *WineBusiness News,* December 10, 2018. https://www.winebusiness.com/news/article/207060.

Trujillo, Marco. "Heatwaves Force Early Spanish Wine Harvests, Nighttime Picking." Reuters, August 30, 2022. https://www.reuters.com/world/europe/heatwaves-force-early-spanish-wine-harvests-nighttime-picking-2022-08-30/.

Turley Wine Cellars. "Bechthold Vineyard Cinsault: Lodi." Accessed October 16, 2023. https://www.turleywinecellars.com/vinesandwines/bechthold-vineyard-cinsault.

VDP. "VDP. Classification." Accessed February 29, 2024. https://www.vdp.de/en/the-wines/classification.

Vins de Bourgogne. "The Birth of Bourgogne Wines: Thousands of Years of History." Accessed February 29, 2024. https://www.bourgogne-wines.com/wine-and-terroir/our-natural-assets/geology/the-birth-of-bourgogne-wines/the-birth-of-bourgogne-wines-thousands-of-years-of-history,2479,9399.html?.

Water Science School, "pH Scale." U.S. Geological Survey, June 19, 2019. https://www.usgs.gov/media/images/ph-scale-0.

Wilson, Chris. "What Is Malolactic Fermentation?—Ask Decanter." *Decanter,* September 15, 2020. https://www.decanter.com/learn/what-is-malolactic-fermentation-51591.

Wine Folly. "The 18 Noble Grapes Wine Challenge." Accessed October 16, 2023. https://winefolly.com/lifestyle/the-18-noble-grapes-wine-challenge/.

Wine Folly. "All About Sekt: Sparkling Wines from Germany and Austria." Accessed February 29, 2024. https://winefolly.com/deep-dive/new-sekt-wine-guide/.

Wines of Germany. "Classification & Labeling." Accessed February 29, 2024. https://germanwineusa.com/basics/classifications/.

Wines of South Africa. "Swartland Independent Producers." September 4, 2012. https://www.wosa.co.za/WOSA-News/Blogs/Cape-Chatter/Swartland-Independent-Producers/.

Wood, Matt. "The Microbiome of Terroir: The Bacteria That Shape the Unique Taste of Your Wine." The University of Chicago Medical Center, April 2, 2015. https://www.uchicagomedicine.org/forefront/biological-sciences-articles/the-microbiome-of-terroir-the-bacteria-that-shape-the-unique-taste-of-your-wine.

ACKNOWLEDGMENTS

Thank you to my editor, Kelly Snowden. I could not have dreamed of a more talented, thoughtful, and especially patient partner in writing this book. Thank you for your vision and diplomacy, your sharp edits and gentle nudges, and for steering me back on course when I'd lost my way. And to Rica Allannic, thank you for encouraging this manuscript and believing I could write it long before I did. I'm sorry it took so long! You're both the best of the best; thank you for choosing to work with me.

Many thanks to the exceptional team at Ten Speed Press, especially Lizzie Allen, whose vision and design brought this book to life; and to Sohayla Farman and Gabby Ureña Matos, whose astute edits and revelatory insights shaped this project in large and small ways. To Kate Bolen, thank you for bringing fresh eyes and salient perspective when the manuscript needed it most.

To Jenny Bowers, your imagination is limitless, thank you for creating gorgeous, whimsical illustrations that transformed these pages beyond my wildest dreams.

This book is based upon a collection of learnings and experiences forged over nearly two decades in wineries, restaurants, and vineyards across the globe. Thank you to all of the wildly talented chefs, winemakers, sommeliers, restaurateurs, and collectors I've had the privilege of working with and learning from during those formative years. Thank you to Jean-Pierre de Smet, who hired me for that inaugural harvest in Burgundy which paved the way for so many more experiences with wine. Thank you to Diana Snowden Seysses and Jeremy Seysses, who let me work the harvest at Domaine Dujac soon after and humored my endless questions (and continue to do so to this day). Thank you to Guillaume d'Angerville, Marie-Andrée and Marie-Christine Mugneret, Jean-Marc Roulot, Alix de Montille, Christophe Roumier, Elena Penna and Luca Currado, Dominique Lafon, and Erin Cannon Chave and Jean-Louis Chave—and so many others—for opening your cellars, for countless remarkable meals, and for many impromptu history lessons that will never all make it into books. Thank you to the late Becky Wasserman, who opened doors for me as she did for so many others, and who lived the example of a life surrounded by family and friends, books, excellent butter, and soulful, delicious wines.

Thank you to Daniel Boulud, who gave me a chance in his kitchen all those years ago and then encouraged me to leave it for wine. Chef, my gratitude is endless. Thank you to Danny Meyer for offering me a job at Eleven Madison Park, which changed the course of my life yet again. Many thanks to John Ragan, who led our wine team at Eleven Madison Park with such conviction, precision, and timely dry humor; and to Will Guidara and Daniel Humm for building a culture of creativity and excellence and leading us on a journey to build what became the number one restaurant in the world. Particular thanks to David Chang, who convinced me to stay in the restaurant industry when I'd decided to leave it and gave me a shot overseeing his beverage programs, a dream job I loved to the last minute. Wylie Dufresne, thank you for hiring me at wd~50 in my earliest years, when I had no business working at such a refined place. You showed us all how to write our own rules.

Thank you to dream mentors Rajat Parr, Bobby Stuckey, Paul Roberts, Aldo Sohm, Brett Zimmerman, and Richard Betts for your grace, generosity, and example and for creating so many opportunities and encouraging my curiosity through those early years and beyond. Thank you to Jane Lopes and Pascaline Lepeltier—I'm much smarter for having studied with you both. And Pascaline, thank you for the time and precision in your final reading of this manuscript. It is so much better as a result.

To Jay Fletcher—you are one of the world's great tasters and an even better human—thank you for being an incredible mentor to me and so many others. Thank you to Su Wong Ruiz, an exceptional thought partner at Momofuku Ko who guided everyone in her orbit to push themselves to be better than they had imagined they could. I always try to channel that.

My deepest gratitude to everyone who encouraged and reviewed this book in its various iterations: my sister Kendyl Salcito, Lisa Brennan-Jobs, Gia Vecchio, Katie Wheaton, and Kate Neuhaus, who read an article I had written and reached out to Rica which started the course of this book.

Thank you to Lhamo and Bhea, invaluable partners at home who allowed me to carve time to write, and to Henry and Ronan, who continue to provide the best distractions.

Thank you to my extraordinary parents, for encouraging my interest in writing since grade school and for your endless love and support to this day.

And last but not least, thank you to Robert, who has been my biggest champion and dreamiest drinking partner since we met two decades ago, at the very beginning of this journey in wine.

ABOUT THE CONTRIBUTORS

JORDAN SALCITO is an award-winning sommelier and entrepreneur who feels fortunate to have lived her dream of working in restaurants and wineries for much of her life. She is a wine industry veteran with over a decade of experience as a sommelier at restaurants, including Momofuku, Restaurant Daniel, and Eleven Madison Park (where she was part of the team to win the award for Outstanding Wine Service from the James Beard Foundation). Her wine programs at Momofuku were regularly recognized in the *New York Times*, Eater, and *Food & Wine*, and, among other awards, were named "Most Creative Wine List in the World" by the *World of Fine Wine* magazine.

Long a student of wine, Salcito passed the tasting portion of the Master Sommelier exam on her first try. She worked the harvest at several of Burgundy's most world-renowned wineries, including Domaines Dujac, Comtes Lafon, and Georges Mugneret-Gibourg. While on maternity leave from her role as the Wine Director for David Chang's Momofuku restaurants, Jordan helped pioneer the ready-to-drink beverage category, founding Drink RAMONA, a collection of award-winning organic Italian spritzes made from Italian grapes and Sicilian citrus fruit. She lives in Paris with her husband, Robert, and their two sons, Henry and Ronan.

JENNY BOWERS works both independently and collaboratively across illustration, animation, and installation projects. She studied illustration in Manchester, animation at the Royal College of Art, then worked in film and television as an art director and graphic artist, joining Peepshow illustration collective in 2003. She has been illustrating, art directing, and very occasionally animating ever since. She is based in Gloucestershire, England.

INDEX

255

Ten Speed Press
An imprint of the Crown Publishing Group
A division of Penguin Random House LLC
1745 Broadway
New York, NY 10019
tenspeed.com
penguinrandomhouse.com

Typefaces: Grilli Type's GT Super, Schick Toikka's Chap, and MVB Fonts's Sweet Sans

Library of Congress Cataloging-in-Publication Data is on file with the publisher.

Hardcover ISBN: 978-1-9848-5882-5
eBook ISBN: 978-1-9848-5883-2

Acquiring editor: Kelly Snowden | Project editors: Kelly Snowden and Kate Bolen
Production editor: Sohayla Farman | Assistant editor: Gabriela Ureña Matos
Designer: Lizzie Allen | Production designers: Mari Gill and Faith Hague
Production manager: Jane Chinn
Copyeditors: Carey Jones and Mi Ae Lipe | Proofreaders: Anne Cherry, Mark McCauslin, Bridget Sweet, and Kate Bolen
Indexer: Jay Kreider
Publicist: Jina Stanfill | Marketer: Joey Lozada

Manufactured in China

10 9 8 7 6 5 4 3 2 1

First Edition

The authorized representative in the EU for product safety and compliance is Penguin Random House Ireland, Morrison Chambers, 32 Nassau Street, Dublin D02 YH68, Ireland, https://eu-contact.penguin.ie.